THE PLANETS

Readings from
**SCIENTIFIC
AMERICAN**

THE PLANETS

Selected and Introduced by
Bruce Murray
California Institute of Technology

Foreword by
Carl Sagan
Cornell University

Sponsored by
The Planetary Society
Pasadena, California

W. H. Freeman and Company
New York San Francisco

Most of the SCIENTIFIC AMERICAN articles in *The Planets* are available as separate Offprints. For a complete list of articles now available as Offprints, write to W. H. Freeman and Company, 660 Market Street, San Francisco, California 94104.

Library of Congress Cataloging in Publication Data
Main entry under title:

The Planets.

    Bibliography: p.
    Includes index.
    1. Planets—Addresses, essays, lectures. I. Murray, Bruce C. II. Planetary Society. III. Scientific American.
QB601.9.P55 1983   523.4   82–21067
ISBN 0–7167–1467–1
ISBN 0–7167–1468–X (pbk.)

Printed in the United States of America

2 3 4 5 6 7 8 9 KP 1 0 8 9 8 7 6 5 4 3

CONTENTS

Note on cross-references to SCIENTIFIC AMERICAN *articles:* Articles included in this book are referred to by title and page number; articles not included in this book but available as Offprints are referred to by title and offprint number; articles not included in this book and not available as Offprints are referred to by title and date of publication.

FOREWORD
Carl Sagan

Our ancestors looked up into the night sky, and of the thousands of shimmering points of light they noticed five that seemed out of the ordinary. Unlike all the other stars, these five changed their relative positions through the course of months. They wandered, in a regular but complex pattern, from constellation to constellation. It was hard to tell what these wandering stars—or indeed the other so-called "fixed" stars—really were. It must have been a topic of protracted speculation and debate.

Eventually the names of gods became attached to them: the faint, fast-moving one that was never far from the Sun was named Mercury, after the messenger of the gods; the most brilliant of them was named Venus, after the goddess of beauty; the blood red one was named Mars, after the god of war; the bright yellowish slow-moving one was called Jupiter, after the king of the gods; and the faint slowest-moving of the five was named Saturn, after the god of time. These metaphorical allusions were the best our ancestors could do: They had no scientific instruments beyond the naked eye, and they were confined to Earth. But in the five centuries since the time of Copernicus, there has been a revolution of historic proportions in our understanding of the nature of these wandering points of light that we now call planets.

We have discovered that every one of them is a world. Not one of them is closely similar to the Earth. We have found dozens of other planets and moons, thousands of asteroids and comets. Since the advent of successful interplanetary flight in 1962, we have flown by, orbited, or landed on more than forty new worlds. We have discovered vast volcanic eminences that dwarf the highest mountain on Earth; ancient river valleys on a planet now too cold for running water; ice worlds that have enigmatically melted; a cloud-covered planet with an atmosphere of corrosive acids and a surface temperature above the melting point of lead; uneroded surfaces preserving some of the history of the formation of the solar system over four billion years ago; exquisitely patterned ring systems, revealing the subtle harmonies of gravity; and a world surrounded by an impenetrable cloud of complex organic molecules like those that in the earliest history of our planet led to the origin of life.

We have uncovered wonders undreamed of by our ancestors who speculated on the nature of those wandering points of light in the night sky. We have begun to probe the mysteries of the origins of our planet and ourselves. By examining other worlds, by discovering what else is possible, by coming face to face with the alternative fates of worlds more or less like ours, we are beginning to understand better our own world. The unmanned exploration of the solar system initiated by the United States and the Soviet Union is a scientific adventure of historic proportions that will be remembered by our remote descendants after much else of our epoch has long been forgotten—provided that we are not so foolish as to destroy ourselves.

But we are just beginning. We have not even completed the preliminary reconnaissance of the solar system. We are able, at a cost that is tiny compared to what we spend regularly on the instruments of mass destruction, to carry out future missions of breathtaking promise—roving vehicles wandering across the ancient Martian landscape; entry probes to sample the rich organic matter of Titan; or manned missions to asteroids whose orbits take them by accident close to the Earth. In this book planetary scientists who have engaged in the great expeditions of exploration and discovery and who are likely to play a role in future such missions describe the new solar system that is gradually emerging before our astonished eyes.

There is enormous public interest, in every nation, in these great voyages and in the emerging new picture that they bring of our universe and ourselves. To focus that interest, a number of us have formed a nonprofit tax-exempt organization called The Planetary Society (P.O. Box 91687, Pasadena, CA 91109 USA). It is the largest space interest group in the world and the fastest-growing membership organization of any kind in the United States in the last decade. It has members on every continent, and its international presence is growing. Nearly all the authors of this book have contributed their royalties to The Planetary Society as a token of the developing partnership between scientists and the public to encourage future interplanetary missions. If the exploratory pace continues—aided in the near future by the entry of many other nations into such spacecraft adventures—then, by the beginning of the twenty-first century, the prospects for a new kind of human future, both on and off the Earth, may begin to become clear.

Carl Sagan
David Duncan Professor
of Astronomy and Space
Sciences, Cornell University
President, The Planetary Society

THE PLANETS

INTRODUCTION

The decade from 1971 to 1980 was characterized by a phenomenal increase in our knowledge and perception of the solar system. During that Golden Decade, all the planets known to the ancients were visited by Earth's robot explorers. The diverse and voluminous findings radioed back to Earth constitute the raw material for the SCIENTIFIC AMERICAN articles reprinted here.

The origin of the Golden Decade can be traced directly back to the 1957 launch of the first artificial satellite by the USSR, followed in 1961 by Soviet cosmonaut Yuri Gagarin's first orbital flight. In the cold war rivalry between the two nations, the United States felt challenged to a highly visible space race and responded with the Apollo program to place an American on the Moon by the end of the 1960s. Looking back from the 1980s, the Apollo program stands out as one of the great American political, technical, and economic decisions of this century. It demanded performance at the very limit of the nation's capacity and yet was an entirely peaceful, adventuresome, and open use of advanced technology. The United States' planetary program likewise had its origin in competition with the Soviets. But by the end of the 1970s, with Voyager explorations of the Jovian and Saturnian systems, broader scientific and popular interest in the nature of our planetary neighbors overshadowed earlier chauvinistic motivations.

From 1967 to 1972, the transport of soil and rock samples from the Moon to Earth for analysis in chemical laboratories, combined with the enormous amount of remote sensing and surface information also acquired, led to a complete revolution in our knowledge of the Moon. Some of the books listed in the General Bibliography summarize the post-Apollo views of the Moon. Thus, as the Golden Decade began, the Moon was already very well known, and it was the primary basis for comparison with other planetary surfaces. A particularly significant finding was that the heavily cratered lunar surface so familiar to Earthbound viewers was not a topographic remnant of the formation of the Moon by accretion, contrary to the general expectations. Instead, the topography records the end of a period of intense bombardment that occurred 500 million years later. This lunar finding implied that Earth also suffered an (otherwise completely unknown) early bombardment about the time the atmosphere and hydrosphere developed and just prior to the time when the earliest rocks still extant on its surface were formed.

Thus, investigation of the Moon led scientists to discover a phase of terrestrial history forever hidden from us on the surface of the Earth. This new field of *comparative planetology* has already contributed other major insights into the origin and evolution of our planet.

In 1974, Mariner 10 carried out the first (and only) exploration of Mercury. It was discovered to be an unexpectedly complex planet. Its surface very strongly resembles that of the Moon—even to the details of the sequence of bombardments recorded there. Yet, incongruously, Mariner 10 discovered a Mercurian magnetic field apparently produced by an internal metallic core. In this respect, it appears that Mercury is more similar to Earth than are any of the other planets. These results are described in the first article, "Mercury."

Venus is nearly Earth's twin in size, mass, and overall chemical composition. But an extensive series of Soviet and U.S. probes has revealed the atmospheric pressure to be almost 100 times greater than that on Earth. Gerald Schubert and Curt Covey's article "The Atmosphere of Venus" describes and elucidates Venus' dense, hot, carbon-dioxide atmosphere with its sulphuric acid clouds and other alien cloud forms. In some ways Venus might represent the bizarre extreme toward which Earth's climate is changing because of our burning of fossil fuels.

Venus' surface has been probed from Earth and from the Pioneer Venus spacecraft by high-resolution radar and at very close range by cameras and instruments aboard Soviet landers. In "The Surface of Venus" by Gordon Pettengill, Donald Campbell, and Harold Masursky, the recorded surface features are shown to exhibit volcanism and some other characteristic terrestrial features in some areas and ancient lunarlike cratered terrains in other areas. These preliminary results suggest that Venus has not developed the full plate tectonics system that so dominates Earth's surface topography and crust.

As described in "The Atmosphere of Mars" by Conway Leovy, Mars exhibits a complete atmospheric contrast to Venus. Its surface pressure is only about one-hundredth that of Earth's. The absence of oceans on Mars and its Earthlike rotation rate make the solar-heat-driven Martian atmospheric circulation simpler to analyze than Earth's atmosphere. Furthermore, the planet is necessarily very sensitive to slight fluctuations in global solar heating, which leads to prominent cyclic climatic variations recorded as layered terrains in the polar areas.

The large and unusual Martian surface features are described in "The Surface of Mars" by Raymond Arvidson, Alan Binder, and Kenneth Jones. For billions of years, there has been extensive igneous activity on Mars as well as conspicuous modification of the surface by wind and floods. These processes have formed gigantic surface features different from those more ephemeral ones on Earth's mobile, highly erosive surface. On Mars, the relatively thick, strong, solid outer layers have effectively precluded plate tectonic activity, which recycles crustal topography on Earth.

"Jupiter and Saturn" by Andrew Ingersoll introduces the giant gas planets, which have been so stunningly revealed by Voyager 1 and 2 following the earlier reconnaissance by Pioneer 10 and 11. Their atmospheric phenomena are in many ways unexpected. Jupiter, especially, manifests large-scale turbulence and high-speed interactions of atmospheric clouds; such interactions might intuitively be expected to gradually mix cloud constituents, resulting in a surface that appears homogeneous. However, large-scale features such as the Great Red Spot have persisted there for at least three centuries!

The next three articles deal with the extraordinary variety of phenomena manifested by the satellites of Jupiter and Saturn, ranging from the sulphur volcanoes of Io to the cold methane and nitrogen atmosphere of Titan and the many icy satellites surrounding both planets. Jupiter and Saturn display extraordinary and continuing interactions between these planetary bodies and their moon systems.

Finally, "Rings in the Solar System" by James Pollack and Jeffrey Cuzzi describes the revolutionary change in our knowledge about both the nature and abundance of rings in the solar system. Until 1976, Saturn was believed to be the only ringed planet, with a beautiful but seemingly simple tripartite ring structure. By 1980, a faint ring had been discovered at Jupiter and rather complex rings had been found at Uranus. Similar features are suspected at Neptune. The Voyager missions revealed Saturn's rings to be composed of tens of thousands of ringlike elements interacting with satellites and with each other in complicated ways. The elucidation of Saturn's rings is a moving reminder of nature's intrinsic beauty, revealed each time we humans substantially increase our observational capacity through the use of new technology.

Following the last article, I provide a brief discussion of anticipated space exploration during the 1980s, including the first look at Uranus and perhaps Neptune by Voyager 2, the in-depth study of Jupiter by the Galileo mission, and the investigations of Comet Halley by spacecraft from many nations. Finally, this collection of articles from SCIENTIFIC AMERICAN, made possible through sponsorship of The Planetary Society, concludes with some personal speculations on where planetary exploration may lead during the next century and beyond.

The solar system. (From the <u>National Geographic Picture Atlas of Our Universe</u>, by Roy Gallant, copyright 1980 by the National Geographic Society.)

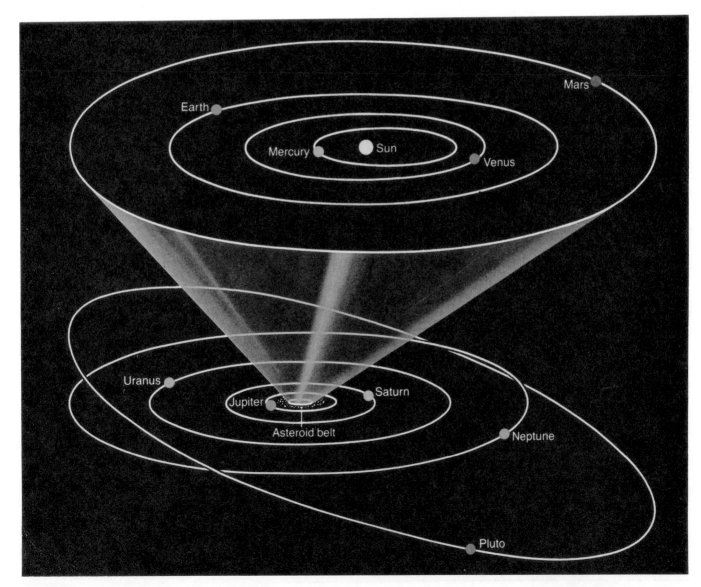

Mercury

by Bruce C. Murray
September, 1975

*The remarkable pictures made by the spacecraft
Mariner 10 have revealed a planetary paradox:
Although Mercury is like the earth on the
inside, it is like the moon on the outside*

There is a story that as Copernicus lay on his deathbed he lamented never having seen the planet Mercury. The story seems implausible because in northern Europe the planet is occasionally visible at twilight. Even if Copernicus could have viewed Mercury through a modern telescope, however, he would have been presented with a singularly unrewarding image. Only a few vague markings can be discerned through a telescope. They are so faint that optical astronomers were long misled into assigning the planet an incorrect rate of rotation.

Last year, 501 years after the founder of modern astronomy was born, the spacecraft *Mariner 10* passed within a few hundred kilometers of Mercury, providing both a 5,000-fold increase in photographic resolution of the planet's surface features and entirely new measurements of phenomena in the planet's immediate environment. Suddenly Mercury was plucked from obscurity and placed in an observational status comparable to that of the moon before the modern age of space exploration. Furthermore, *Mariner 10* is in an orbit that carries it back to the vicinity of Mercury every 176 days. As a result the spacecraft transmitted a second set of close-up photographs on September 21, 1974, 176 days after its first encounter with the planet, and an extremely valuable third set of observations, including a limited set of high-resolution pictures, on March 16 of this year, shortly before it exhausted its supply of gas for stabilizing its attitude in space.

Before *Mariner 10*'s highly successful voyage it was recognized that Mercury is covered with at least a thin layer of finely divided dark silicate material very similar to that on the moon. Allowing for the difference in distance from the sun, Mercury closely mimics the moon in the way it reflects sunlight and radar pulses and in its emission of infrared radiation and radio waves. Yet it has been known on the basis of Mercury's size and mass that the planet is much denser than the moon or Mars and only slightly less dense than the earth. The earth's bulk density is greater than the laboratory density of its constituent materials, since much of the earth's substance is under high pressure in its interior. Hence the high bulk density of the much less massive Mercury implies that Mercury contains an even greater abundance than the earth of heavy elements, particularly iron.

Even the two most elementary facts about Mercury inferred from ground-based observations—the nature of its surface materials and its density—raised difficult questions. Could Mercury be composed of a homogeneous mixture of iron and silicate materials, as is the case with certain kinds of meteorites? Alternatively, could Mercury have a large earthlike iron core enclosed by a relatively thin silicate mantle and crust? If Mercury is chemically differentiated as the earth is, and the evidence now points in that direction, the diameter of its iron core is fully three-fourths the diameter of the planet, or the size of the moon!

Since Mercury is the innermost planet in the solar system, never more than 28 angular degrees from the sun in the sky, it is notoriously difficult to study by conventional astronomical techniques. We now appreciate how seriously these techniques can be compromised by contamination with sunlight. (Mars, in comparison, presents its largest image in the middle of the night, when the earth lies directly between it and the sun.) As recently as 1962 a leading expert in the visual and photographic observation of the planets wrote in what was then the most authoritative book on planetary astronomy that Mercury rotates at a rate such that one hemisphere constantly faces the sun, as one hemisphere of the moon constantly faces the earth. This synchronous rotation meant that Mercury supposedly turned on its axis once every 88 days, in step with the period of its revolution around the sun. Indeed, the planet was said to be synchronous with the sun to within one part in 10,000. In the same period spectroscopists working at two different wavelengths concluded that Mercury has a thin atmosphere. That conclusion received independent support from purported variations across the disk of the planet in the degree to which reflected sunlight

CRATERED SURFACE OF MERCURY was photographed for the first time late in March of last year by cameras aboard *Mariner 10*. This high-resolution picture shows a typical heavily cratered region, strongly resembling the surface of the moon, on the equator of the planet. The pictures returned by *Mariner 10* made it possible to relate a previously agreed-on longitude system for Mercury to a specific topographical feature for the purposes of detailed mapping. In 1970 the International Astronomical Union had defined the origin of planetographic longitudes as the meridian crossing the subsolar point of the first perihelion passage of 1950. It has now been agreed that the 20-degree meridian of Mercury passes through the center of a particular small crater immediately adjacent to the large crater near the center of this picture. The small crater, which is 1.5 kilometers in diameter and lies .58 degree south of the equator, is at the foot of the outer rim of the large crater at a position equivalent to eight o'clock on a clock dial. It has been named Hun Kal, which stands for 20 in the language of the Maya, who used a base-20 number system. The photograph is reproduced with north at right in order to include as much of *Mariner 10* frame as possible.

was plane-polarized. We now know that Mercury has no atmosphere whatever, and has not had one for billions of years. And the planet does not revolve synchronously around the sun.

On the other hand, imaginative artists had commonly depicted the surface of Mercury as resembling the surface of the moon, and their intuition has proved to be correct. The *Mariner 10* photographs reveal that Mercury's surface is remarkably similar to the moon's, not only in its features but also in the sequence of events that was required to produce them. Mercury thus presents something of a planetary paradox: it is like the moon on the outside, yet like the earth on the inside, even to exhibiting an earthlike magnetic field.

In 1962, after centuries of unsatisfactory observations at visible wavelengths, radio waves from Mercury were detected. Radio astronomers from the University of Michigan observed the planet near elongation, when half of its disk, as it is seen from the earth, is in sunlight and half is in shadow. If Mercury were in synchronous rotation around the sun, the dark side would never receive any direct radiation from the sun and would be perpetually cold. Hence the thermal emission (including radio waves) from the dark side should be extremely low, well below the limit of detection. The Michigan workers were surprised to discover a substantial total radio flux evidently arising from both the dark and the sunlit halves of the planet, corresponding to an overall average near-surface temperature of 350 to 400 degrees Kelvin (170 to 260 degrees Fahrenheit). Such an average apparent temperature is exactly what any moonlike object would exhibit at the orbit of Mercury if it were rotating on its axis more rapidly than the once-a-revolution synchronous rate. Astronomers were so committed to the idea of synchronous rotation, however, that it was generally presumed that the anom-

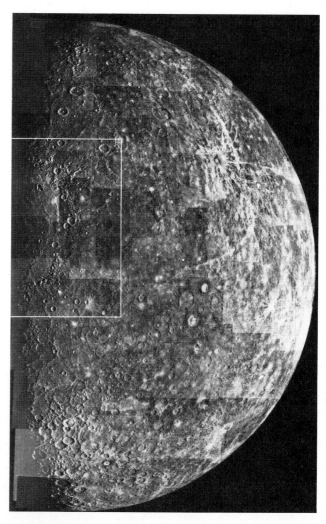

TWO HEMISPHERES OF MERCURY, each approximately half in shadow, were photographed during *Mariner 10*'s first encounter with the planet in March of last year. The mosaic of high-resolution pictures at the left shows the "incoming" hemisphere: the hemisphere visible as the spacecraft approached the planet, before sweeping behind its dark side. The evening terminator, the shadow line at the right, lies near 10 degrees west longitude. Since *Mariner 10* approached Mercury from below the plane of its orbit, the center of the disk in the view at the left is about 20 degrees south of the equator. The area within the upper rectangle, which encloses the bright-rayed Kuiper Crater, is shown enlarged in the illustrations on the opposite page. The area within the lower rectangle appears enlarged on page 8. The mosaic at the right shows the "outgoing" hemisphere of Mercury, the hemisphere visible as *Mariner 10* "looked back" after passing behind the dark side of the planet. The spacecraft is now viewing Mercury from a point about 20 degrees north of the equator. The morning terminator, the shadow line at the left, lies near 190 degrees west longitude. The large impact structure named the Caloris Basin, comparable to the Imbrium Basin on the moon, is half-visible on the terminator just north of the center of the disk. The region inside the rectangle is shown on pages 42 and 43 in a sequence of pictures of increasing resolution.

alous thermal emission from the dark half must indicate the presence of an atmosphere capable of transporting heat from the day side to the night side.

In 1965 Rolf B. Dyce and Gordon H. Pettengill carefully measured the differences in frequency among returning radar pulses beamed at the edges of Mercury from the Arecibo Observatory. They concluded that the planet did not rotate synchronously around the sun but instead had a rotation period of 59 ± 5 days in the direct sense (in the same sense as the earth's rotation). They did not, however, mention in the scientific publication of their results that this finding would explain the "anomalous" heat emission from the dark side of the planet. Even a year later a comprehensive review article was devoted to the putative atmosphere of Mercury and its presumed role in the transport of heat.

Why a 59-day rotation period? Giuseppe Colombo, an Italian dynamicist with a long interest in Mercury, quickly recognized that a period of 59 days stood in relation to the 88-day period of the Mercurian year about in the ratio 2 : 3. Colombo conjectured that Mercury's rotation period is, in fact, precisely 58.65 days, which means that the planet would rotate exactly three times while circling the sun twice, thereby exhibiting the phenomenon of spin-orbit coupling. The conjecture has been fully confirmed not only by further radar observations but also by the photographs from *Mariner 10*.

It is extremely improbable that Mercury exhibits spin-orbit coupling simply by coincidence. It is more likely that tidal interaction with the sun has removed angular momentum and slowed the planet sufficiently from a higher original rate of spin to trap it into the present resonant period. Such a theory was quickly developed by Peter Goldreich and Stanton J. Peale and by Colombo and Irwin I. Shapiro.

The success of the *Mariner 10* mission depended on a three-body interaction calling for an assist from Venus. The spacecraft initially followed a course that took it close to the intervening planet, where its trajectory was perturbed toward the orbit of Mercury. In this way an entirely different orbit around the sun was achieved, an orbit otherwise quite unobtainable with a launch vehicle of the Mariner class. The new orbit carried *Mariner 10*, nearly at perihelion (the point closest to the sun), to an encounter with Mercury when the planet, traveling in its own eccentric orbit, was nearly at aphelion (the point farthest from the sun). An initially unforeseen property of

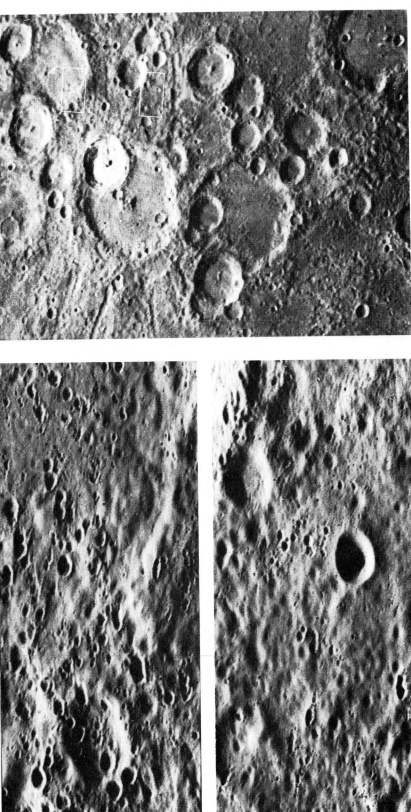

KUIPER CRATER, named for the late Gerard P. Kuiper, who helped to plan the *Mariner 10* photography of Mercury, is shown in the general view (*top*) taken during the first encounter. The white rectangles identify the location of two high-resolution frames (*bottom*) taken during *Mariner 10*'s third encounter with the planet on March 16 of this year. Kuiper Crater, roughly 40 kilometers in diameter, reflects sunlight more strongly than any other feature observed on Mercury. The picture at the top has been computer-enhanced to bring out small-scale features while suppressing most of the variations in reflectivity. The two high-resolution frames show streams of secondary craters produced by material ejected by the impact that created Kuiper Crater. Resolution in these two frames is about 250 meters.

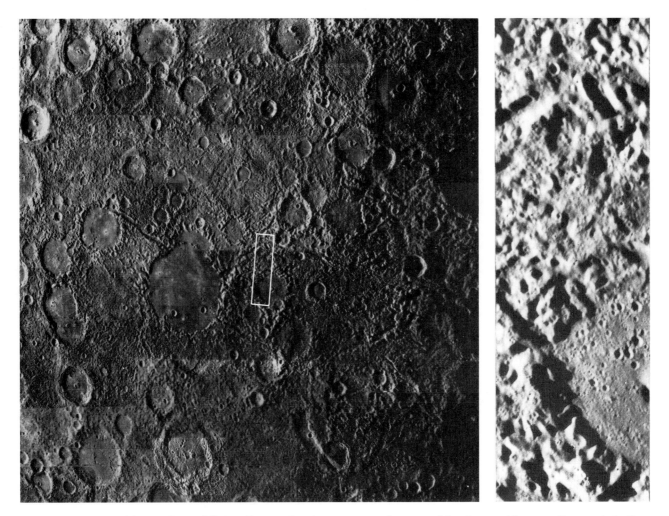

PECULIAR TERRAIN lies southeast of Kuiper Crater and antipodal to the Caloris Basin. The area shown in the picture at the left lies within the lower rectangle in the mosaic at the left on page 38. One hypothesis is that the hilly and lineated terrain shown here was created by the seismic effects of the great impact that created the Caloris Basin on the opposite side of the planet. The large crater near the center of the picture, with two small craters in its floor, is about 170 kilometers in diameter. The area within the white rectangle is shown in the high-resolution picture at the right, which was taken during *Mariner 10*'s third encounter with the planet. The picture resolves surface features as small as 450 meters. The half-visible large crater is approximately 55 kilometers in diameter.

this exquisite celestial billiard shot is that *Mariner 10*'s new orbit has a period exactly twice that of Mercury's. As a result *Mariner 10* will continue to return every two Mercurian years to pass Mercury at exactly the same heliocentric longitude. Since Mercury itself spins precisely three times on its own axis in two Mercurian years, the spatial orientation and surface illumination of the planet at each encounter with *Mariner 10* will be exactly the same [*see top illustration on page 12*]. Thus Mercury, the sun and the spacecraft are dynamically in a state of triple resonance.

Mercury has the harshest surface environment of any planet in the solar system. When Mercury is at perihelion, it receives 10 times as much solar energy per unit of surface area as the moon. Noontime temperatures at the equator of Mercury soar to 700 degrees

K., and in the dark hemisphere the surface cools radiatively to less than 100 degrees. Furthermore, "noontime" lasts a long time at Mercury's perihelion because of the coupling of the planet's rotation with the period of its eccentric orbit. An observer on Mercury at perihelion would see the sun slow to a complete halt in its motion across the sky and then move slightly in a retrograde direction (westward through the constellations) for eight days [*see bottom illustration on page 12*]. The reason is that the orbital angular velocity exceeds the spin angular velocity near perihelion. Moreover, the areas subjected to the longer period of solar radiation at perihelion are always near the same longitudes: 0 degrees and 180 degrees. The longitudes 90 degrees and 270 degrees receive their maximum solar irradiation at aphelion. As a result the 0-degree and 180-degree meridians receive two and

a half times more solar radiation overall than the meridians 90 degrees away from them. Hence even though Mercury's spin axis is probably perpendicular to the plane of the planet's orbit, so that there would be no seasonal variations with latitude as there are on the earth and on Mars, the spin-orbit coupling of Mercury gives rise to a seasonal variation in temperature with longitude.

Another interesting property of Mercury is that in its equatorial regions the subsurface temperatures are always above the freezing point of water and in the polar regions the subsurface temperatures are well below freezing. In contrast, liquid water of internal origin cannot reach the surface of the moon or Mars (except through volcanic activity) because the subsurface temperatures are everywhere below freezing for a depth of many kilometers. In view of the large longitudinal variation in the influx of solar radiation and the potential for a latitudinal variation in chemical weathering conceivably associated with the occasional release of subsurface water, one might expect to perceive some characteristic effects on the surface at appropriate geographic locations on the planet. Actually we have been surprised to find no such effects either in radar maps made from the earth or in the photographs sent back by *Mariner 10*.

The illuminated surface observed by *Mariner 10* as it first approached Mercury is dominated by craters and basins, creating a landscape that could easily be mistaken for parts of the moon. There are, however, significant differences. The heavily cratered regions of Mercury exhibit conspicuous plains, or relatively smooth areas, between the craters and the basins, whereas the highlands of the moon generally show densely packed and overlapping craters. The intercrater plains appear in many cases to predate the formation of most of Mercury's large impact craters. The surface of Mercury is also unlike the surface of the moon in that it is not saturated with large craters having diameters between 20 and 50 kilometers.

One factor contributing to the difference in surface appearance between Mercury and the moon has been suggested by Donald E. Gault of the National Aeronautics and Space Administration. He points out that since on the surface of Mercury the force of gravity is twice that on the moon, material ejected from a primary impact crater on Mercury will cover an area only a sixth as large as the area covered on the moon for an impact crater of the same size. On Mercury secondary impact craters are much more closely clustered around primary craters than they are on the moon. As a result the topographic record of early events may be better preserved on Mercury than it is on the moon, where ejecta from the most recent impact basins has blanketed much of the earlier surface.

Another important difference between the heavily cratered regions of Mercury and those of the moon is the ubiquitous presence on Mercury of shallowly scalloped cliffs running for hundreds of kilometers. The structure of these features, termed lobate scarps, suggests they resulted from an early period of crustal shortening on a global scale. Such features are not conspicuous on the moon or on Mars. On the contrary, on those two lower-density bodies one sees tectonic evidence of crustal stretching. Robert G. Strom of the University of

VIEWS TAKEN 176 DAYS APART by *Mariner 10* on its first encounter (*top*) and on its second encounter (*bottom*) show that the angle of solar illumination in each case is virtually identical, thereby supporting earlier evidence that Mercury rotates on its axis exactly three times while going around the sun twice. The planet's orbital period is 88 days and its rotation period is 58.65 days. Bottom picture shows a smaller area and is more distorted than top picture because it was taken closer to the planet and at a more oblique angle.

Arizona and others have speculated that the lobate scarps of Mercury are the result of a long period of crustal shortening produced by the slow cooling and contraction of Mercury's large iron core. In any case the very existence of large, well-preserved craters on Mercury, which are probably three or four billion years old, is evidence that there has been no planetwide melting or earthlike migration of crustal plates since that time. Furthermore, the evident lack of surface erosion rules out any tangible atmosphere for Mercury as far back as the time when the craters were made. In contrast the surface of Mars illustrates clearly how even a tenuous atmosphere quickly blankets and modifies the appearance of large craters, notably removing the initially conspicuous bright "rays" of ejecta that radiate from them.

While the television cameras of *Mariner 10* were revealing the ancient cratered terrain of Mercury, the magnetometer, the plasma probe and the charged-particle detector aboard the spacecraft were recording an interaction with the "wind" of charged particles from the sun that was much stronger than had been anticipated. The track of *Mariner 10* had been calculated so that the spacecraft would pass the dark side of the planet on its initial encounter; in that way the instruments would be able to investigate the wake left by Mercury in the solar wind [*see bottom illustration on page 13*]. A close passage in the wake of a planet provides the least ambiguous observational path for a flyby, particularly to test for the presence of anything resembling an earthlike magnetic field.

On the first flyby, when *Mariner 10* passed about 700 kilometers above the surface of Mercury, the appropriate instruments detected a weak magnetic field and a generally earthlike interaction with the solar wind. In order to get more definitive observations the spacecraft was targeted to make its third pass still closer to the surface (327 kilometers) on a track that carried it closer to the north pole. The third flyby confirmed the strength and orientation of the magnetic field that had been predicted by Norman F. Ness of NASA on the basis of the first flyby. Mercury evidently has a dipole magnetic field approximately aligned with the planet's spin axis. The strength of the field ranges from 350 to 700 gammas at the surface, or about 1 percent of the strength of the earth's surface magnetic field. Mercury's field is far stronger than the field found at either Venus or Mars, and a planetwide internal mechanism seems to be required for its generation. In addition Mercury is surrounded by a very thin envelope of helium gas, suggesting that the planet's magnetic field entraps helium nuclei from the solar wind and possibly from surface emanations as well.

The existence of a seemingly earthlike magnetic field on Mercury certainly provides independent evidence for a chemically differentiated planet with an iron core. Pictures acquired on the second

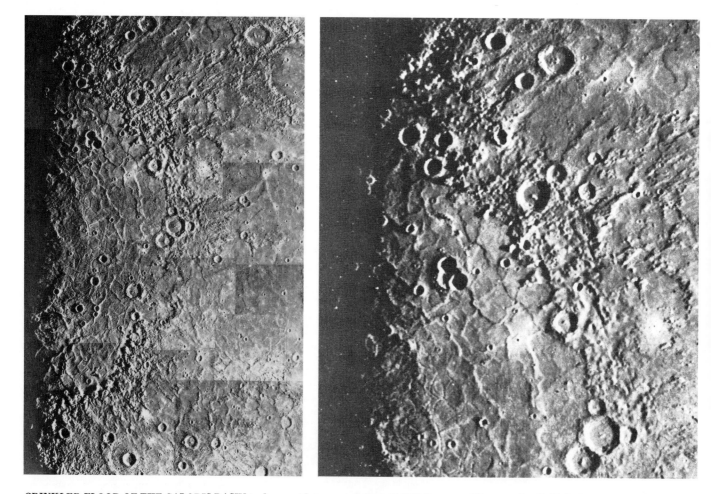

CRINKLED FLOOR OF THE CALORIS BASIN is shown with increasing resolution in the sequence of five pictures across these two pages. The first picture is a mosaic of views taken during *Mariner 10*'s first encounter with Mercury, looking back from a distance of about 75,000 kilometers. The second and third pictures were taken earlier in the same pass, when the spacecraft was closer to the planet. The fourth and fifth pictures were taken 352 days later, when the spacecraft made its third close approach to Mercury after

flyby, last September, showed that the lobate scarps seen in the first series of pictures continue across the south-polar region. Pictures from the first and second passes also show that small, steep-walled craters near the two poles have areas that are permanently shaded from the sun and thus constitute enduring "cold traps" for whatever volatile substances may have been released intermittently over Mercury's history. Those sites might be an exciting objective for close inspection in the 21st century.

The "outgoing" hemisphere of the planet photographed by *Mariner 10* as it "looked back" after its close approach exhibits a configuration of surface features totally different from those presented by the "incoming" hemisphere [*see illustration on page 6*]. The outgoing hemisphere shows large areas of relatively smooth plains that are clearly younger than most of the heavily cratered terrain visible in the incoming hemisphere. In addition there is a 1,400-kilometer basin left from a gigantic impact comparable to the one that gave rise to the Imbrium Basin on the moon. This prominent Mercurian feature, named the Caloris Basin because of its equatorial location near the "hot" 180-degree longitude of Mercury, is, like the Imbrium Basin, entirely filled with smooth plains material.

Other plains, less cratered and deformed, extend eastward and northward for thousands of kilometers. The *Mariner 10* television team has concluded that the temporal sequence of these plains, their variation in reflectivity and color, their size relations and their geographic association all point to a widespread episode of volcanism that followed the end of the period of heavy bombardment. The resolution provided by the pictures does not, however, reveal the surface morphology clearly enough to unambiguously identify the origin of the plains.

The situation is reminiscent of the debate about the origin of the smooth plains on the moon before the Apollo missions. It appears that some of the lunar plains were created by gigantic impacts rather than by volcanism. Hence the possibility remains that the material covering the Mercurian plains consists of massive sheets of ejecta from huge impacts, conceivably located on the hemisphere of Mercury that was in shadow during the three *Mariner 10* encounters. Whatever the origin of the plains, the fact that they have not been crumpled by internal activity and have not been modified by atmospheric erosion or deposition testifies to the remarkable quiescence of Mercury since the plains were made. Indeed, the overall similarity in the sequence of events that shaped the surface of Mercury and the moon is extraordinary and, to me, surprising, considering how very different the internal constitution of the two bodies must be.

A preliminary surface history of Mercury can be divided into five major sequences of events, each generally similar to those that shaped the moon. First,

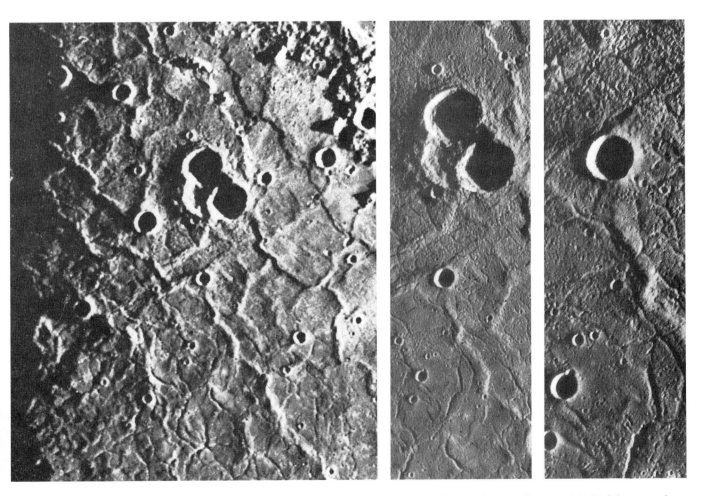

completing two trips around the sun on its own 176-day orbit. The fractures that are visible in the floor of the Caloris Basin vary in width from about eight kilometers down to about 450 meters, the limit of photographic resolution. The largest crater in the fifth picture, which was taken at a distance of 20,000 kilometers above the surface of the planet, is about 12.5 kilometers in diameter. Each of the two young craters in the prominent cluster of three craters that is seen in the other pictures is about 35 kilometers in diameter.

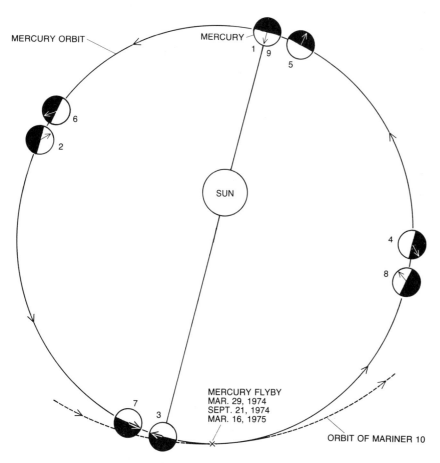

PHENOMENON OF SPIN-ORBIT COUPLING is what has locked Mercury's rotation period and orbital period in the ratio of two to three. In this diagram of Mercury's orbit the fixed arrow points toward one of the planet's two hot subsolar points, that is, the points on the equator that lie directly under the sun at alternate perihelions. The numbers give the sequential position of the planet in its orbit during two of its revolutions around the sun. *Mariner 10*'s encounters with Mercury take place at the point that is marked with an **X**.

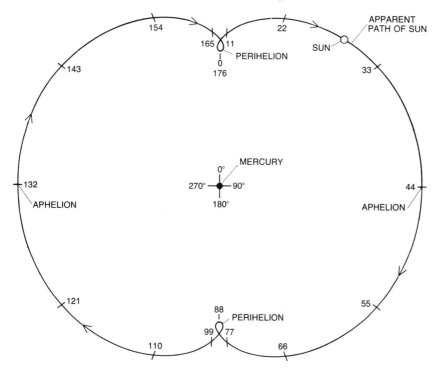

SUN AS SEEN FROM MERCURY appears to execute a loop at perihelion. Apparent position of the sun in relation to subsolar longitudes on the planet is marked off in 11-day intervals for two Mercurian years. Pattern of motion was described by S. Soter and J. Ulrichs.

the absence of recognizable volcanic, tectonic or atmospheric modification of the large old craters on Mercury implies that the mass of the planet was chemically fractionated into a large iron core enclosed by a thin silicate mantle well before any of the oldest craters were formed. Any atmosphere on Mercury must have escaped quite early or was never formed. The heat required for the chemical separation of iron and silicate phases on a global scale must also have dissipated early enough for the outer layers to become sufficiently rigid to maintain the topographic relief of the large old impact craters up to the present time.

In the second major epoch, following the initial period of accumulation and chemical segregation, Mercury must have gone through at least one period of crater obliteration, possibly through early volcanism, if we are to account for the absence of topographic scars of the accretion process. The smoothed surface, surviving in the plains between craters, records not only the terminal phase of heavy bombardment but also the global shrinking of the crust represented by the lobate scarps.

The third and most sharply delineated of the five major subdivisions of Mercury's surface history came toward the end of the heavy bombardment with the large Caloris Basin impact, seemingly the counterpart of the lunar impact that created the Imbrium Basin. The Mercurian collision gave rise both to the Caloris Basin and to the mountainous terrain surrounding it and extended areas of ejecta and sculpturing of the older surface. In addition, approximately antipodal to the Caloris Basin there is a peculiar lineated and hilly terrain, which Gault and Peter Schultz speculate may have been thrown up by the focusing of seismic energy from the Caloris Basin impact on the opposite side of the planet.

During the fourth phase, some time after the Caloris Basin impact, broad plains were created, probably as a result of widespread volcanism similar to the activity that gave rise to the lunar maria, or "seas." Bruce W. Hapke of the University of Pittsburgh argues that the colormetric and photometric evidence from *Mariner 10*, combined with previous telescopic results, suggest that the surface rocks on Mercury may be somewhat less rich in iron and certainly less rich in titanium than the rocks found on the maria of the moon, thus explaining the absence on Mercury of the sharp contrast in brightness levels between the lunar maria and the lunar highlands.

During the fifth phase of the surface history of Mercury, extending to the present time, little has happened other than a light peppering of impacts, many of which show conspicuous rays. The distribution of impact craters on the Mercurian plains is remarkably similar to the distribution both on the maria of the moon and on some of the oldest smooth plains of Mars. The similarity of impact-cratering rates on Mercury, the moon and Mars over the past three billion years or so came as a surprise to many, considering the great differences in the three bodies' distance from the sun and therefore the differences in the probability of their encountering interplanetary debris from the asteroid belt (which has long been thought to be the principal source of the impacting objects).

Thus *Mariner 10* has clearly established that the exterior of Mercury resembles the moon not just in topography but more surprisingly in surface history. And yet the interior constitution of Mercury appears to be more earthlike than that of any of the other planets. The paradoxical circumstance of Mercury's moonlike exterior and earthlike interior raises important questions not only about Mercury but also about the history and nature of the entire inner part of the solar system. Were the bombarding objects whose impacts are recorded on the surface of Mercury from the same family of objects that bombarded the moon as recently as four billion years ago? Or did each of the inner planets, including the moon, pass through separate periods of late bombardment that overlapped only slightly, if they overlapped at all, and in each case ceased abruptly and independently?

The *Mariner 10* pictures suggest to me and to Newell J. Trask, Jr., of the U.S. Geological Survey that the last great bombardment of Mercury, culminating with the Caloris Basin impact, may not have been part of a steadily decreasing flux of interplanetary debris but could have resulted from a discrete terminal episode of bombardment. George W. Wetherill of the University of California at Los Angeles considers it plausible that the bombarding objects involved in such an early discrete episode could have originated with a single object perturbed to pass near the earth or Venus from an initial orbit beyond Jupiter. Tidal disruptions on the earth or Venus might then conceivably have created a shower of bombarding objects that would have been rapidly swept up through collisions with the four inner

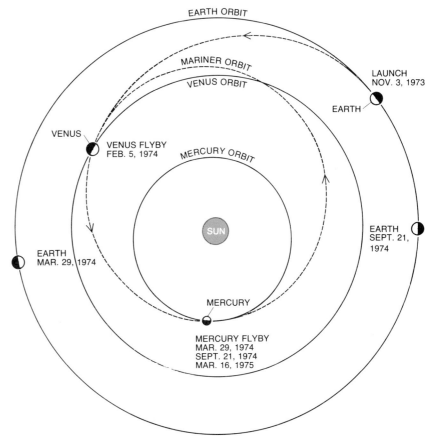

TRAJECTORY OF MARINER 10 was chosen so that the spacecraft was deflected toward Mercury by a precisely timed encounter with Venus three months after leaving the earth. It is as a result of this deflection that the orbit of *Mariner 10* takes it to the vicinity of Mercury every 176 days, or once every two Mercurian years. It was this happy choice of orbital characteristics that enabled *Mariner 10* to achieve three operational encounters with its target.

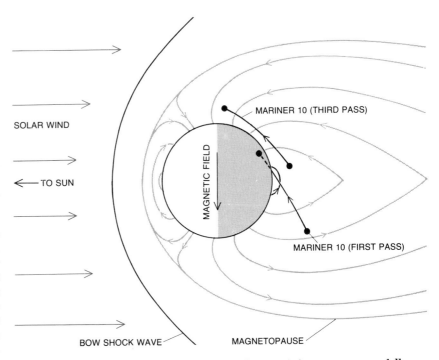

MAGNETIC FIELD OF MERCURY, detected in *Mariner 10*'s first encounter, was fully confirmed in the third encounter, when the spacecraft was targeted to pass closer to the north pole of the planet. The black dots indicate in planar projection when *Mariner 10* entered and left the planet's magnetic field on each pass. Field lines are distorted by pressure of charged particles in the solar wind. The polarity of Mercury's field is oriented like that of the earth.

planets. Such a concept would be consistent with recent controversial proposals that the moon was subjected to a similar terminal episodic bombardment about four billion years ago.

On the other hand, if the observed topography was created by a continuously declining bombardment of Mercury, the evidence may have been abruptly and effectively erased by an episode of enhanced crater obliteration.

Hence the appearance of an episodic bombardment of the planet may be an illusion. The debate now developing over the early history of the inner solar system is reminiscent of an earlier debate between uniformitarians and catastrophists over the causes of the earth's geological features. There the uniformitarians won.

That Mercury has a dipole magnetic field aligned with its spin axis, very similar to the earth's field although weaker, is to me particularly unexpected. Granted that Mercury probably has a large iron core, the rotation of the planet is nevertheless so slow at present that before *Mariner 10*'s encounter with the planet no one thought that a Mercurian field might be generated by a fluid-dynamo mechanism of the type postulated for the earth (in which the field arises from electric currents associated with fluid

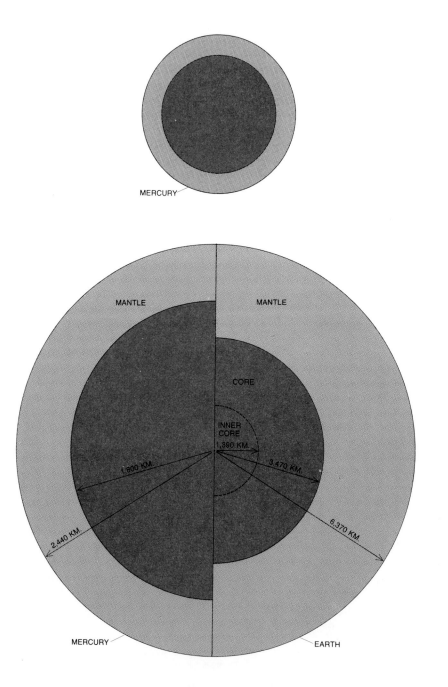

MERCURY

MANTLE   MANTLE

CORE

INNER
CORE
1,390 KM.

1,800 KM.   3,470 KM.

2,440 KM.   6,370 KM.

MERCURY   EARTH

CUTAWAY VIEWS OF MERCURY AND THE EARTH, which are scaled to have the same outside diameter, show how much larger Mercury's iron core is thought to be compared with earth's core. Both of the cores are shown in dark color. Mercury's actual size in relation to the earth is shown by the small disk at the left. Mercury's iron core evidently contains 80 percent of the planet's mass. Therefore the iron core must have a radius of at least 1,800 kilometers, which would make core alone slightly larger than the earth's moon.

motions in the core of the spinning earth). And what about Venus, which presumably has a larger and hotter core than Mercury's and yet does not exhibit a significant planetary magnetic field? Furthermore, if there are fluid motions within the Mercurian core capable of generating the observed magnetic field, the core motions or the associated heat flow have not led to any recognizable deformations of the planet's surface layers.

A quite different explanation for the planet's field is that it is a fossil field of some kind that has persisted from a remote epoch. It seems unlikely, however, that in the billions of years since the hypothetical fossil field was created the temperature within the appropriate por-

tion of Mercury's interior never rose above the Curie point (the temperature at which a substance loses its magnetism). Still a third possibility is that the field has somehow been induced as a result of Mercury's continued interaction with the solar wind. Preliminary examination of this concept suggests that such a field would not exhibit symmetry around the rotation axis.

Perhaps the Mercurian magnetic field arises from causes still unimagined. Or perhaps we shall have to gain a deeper understanding of the mechanism of the earth's own field in order to explain how that mechanism could be reduced to the Mercurian scale. Whatever the outcome, it is fortunate that there is another magnetic field among the inner planets with

which to compare and contrast the field of the earth. Further study of Mercury's field, as well as photographic mapping of the still unobserved hemisphere of the planet, would be a major objective of any orbiting satellite of Mercury.

Mariner 10's mission to Mercury has completed the reconnaissance of the inner solar system and has further demonstrated that planetary exploration is full of surprises. The findings of *Mariner 10*, combined with the testimony of the lunar samples brought back by the Apollo astronauts, have made possible an enormous extension of knowledge and inference about Mercury. Out of the new observations is emerging a richer and more unified picture of the origin and evolution of all the planets, including our own.

2

The Atmosphere of Venus

by Gerald Schubert and Curt Covey
July, 1981

A decade of exploration by spacecraft now shows that it consists almost entirely of carbon dioxide. Its clouds of sulfuric acid are driven by winds that attain a speed of 360 kilometers per hour

Until a few decades ago the atmosphere of Venus was unknown. Three centuries of observations with telescopes had revealed only that the planet is covered by an unbroken deck of seemingly uniform, featureless clouds. Today observations made from the earth have been extended to ultraviolet wavelengths. In this part of the electromagnetic spectrum markings are seen in the clouds. Moreover, since 1967 a series of interplanetary spacecraft have made observations at closer range, and a total of 13 probes have penetrated deep into the atmosphere of Venus before the heat of the atmosphere disabled them. Remarkably, two of the spacecraft in the Venera series launched by the U.S.S.R. survived long enough to return data from the surface, where the year-round temperature of 460 degrees Celsius is high enough to melt metals such as zinc. A U.S. spacecraft, the Pioneer Venus orbiter, which arrived at Venus in December, 1978, continues to circle the planet and transmit information back to the earth.

Today the atmosphere of Venus is known to be predominantly (96 percent) carbon dioxide. The rest is nitrogen with trace amounts of other substances, including water vapor. The atmosphere of Venus is more than 90 times as massive as the atmosphere of the earth, so that the pressure at the surface of Venus is more than 90 times the pressure at the surface of the earth. It is a pressure encountered at a depth of one kilometer in the oceans of the earth. The clouds in the atmosphere of Venus are now known to form continuous layers at an altitude roughly between 45 and 60 kilometers. (On the average the earth is half covered by clouds that lie at an altitude of less than 10 kilometers.) Investigators distinguish three layers of clouds on the basis of the concentration and the size of suspended particles. It is hypothesized that the particles may differ from one layer to another, but the only particles identified so far are liquid droplets composed of sulfuric acid with an admixture of water.

The winds of Venus are now known to be dominated by a planetwide east-to-west circulation that attains a velocity of 100 meters per second (360 kilometers per hour) at the altitude of the cloud tops. At that speed a parcel of the atmosphere would circle Venus in four earth-days. The planet itself also turns east to west, but it takes 243 earth-days to complete a single rotation. In a word, the atmosphere of Venus superrotates: at the cloud tops it moves more than 60 times as fast as the planet does. In contrast the earth turns west to east, and the atmosphere of the earth (considered on a global scale) rotates synchronously with the solid planet below it.

The focus of research on the atmosphere of Venus is now changing. The challenge is no longer to discover what the atmosphere is like. The challenge is to explain why the atmosphere of Venus should be so different from that of the earth in spite of the fact that Venus and the earth are nearly the same in size and mass.

The atmosphere of a planet is well characterized by the way its temperature varies with altitude. On Venus the pattern of variation divides the atmosphere into two regions. In the lower one, which extends from the surface to an altitude of about 100 kilometers, the temperature decreases with height. This region is called the troposphere by analogy with the lower part of the atmosphere of the earth (the part below 10 kilometers), in which the temperature

also decreases. Throughout most of the height of the troposphere of Venus the temperature falls at a rate of about 10 degrees C. for each kilometer of altitude. At the surface the temperature of 460 degrees changes little from day to night. Indeed, one must make measurements at the height of the clouds to find diurnal changes. The reason for this lack of variation is clear. The lower atmosphere is so dense that it has great thermal inertia: it can store a large amount of heat. For the same reason one would not expect to find a significant difference between the daytime and the nighttime temperature in the oceans of the earth.

The high temperatures below the clouds, culminating in the surface temperature of 460 degrees, cannot be attributed simply to the fact that Venus is closer to the sun than the earth is; indeed, the ubiquitous cloud cover on Venus is so reflective that the planet absorbs less solar radiation than the earth. Furthermore, as has been shown by Pioneer Venus probes and by the Venera series of probes, only a small fraction of the radiation absorbed by Venus penetrates the clouds and the massive lower atmosphere and reaches the surface. The planet remains hot in spite of all this because the surface reradiates the solar energy in the infrared part of the electromagnetic spectrum; constituents of the atmosphere such as carbon dioxide, sulfur dioxide, water vapor and cloud particles are efficient absorbers of infrared radiation, and so the heat is trapped in the troposphere. In short, the tropo-

DOMINANT WIND ON VENUS is revealed in the series of photographs on the opposite page, which record the image of the planet in ultraviolet radiation. The photographs were made on consecutive days by the Pioneer Venus orbiter, a U.S. spacecraft that took up its trajectory around the planet in December, 1978. In the first photograph, made on February 15, 1979, a *Y*-shaped marking in the clouds of Venus opens toward the west of the planet. The stem of the *Y* is a broad, dark band along the equator. In the second photograph the vertex of the *Y* is near the west limb of the planet; in the third photograph the vertex is out of view; in the fourth the arms are returning; in the fifth, made on February 19, the complete *Y* is again exposed. The dominant wind is thus a rotation of the atmosphere from east to west that drives cloud markings around the planet in only four earth-days. Venus itself also rotates westward, but the rotation of the solid planet is more than 60 times slower than that of the atmospheric markings. The final photograph shows Venus on February 20, 1979. The photographs were provided by Larry D. Travis and Anthony Del Genio of the Goddard Institute for Space Studies.

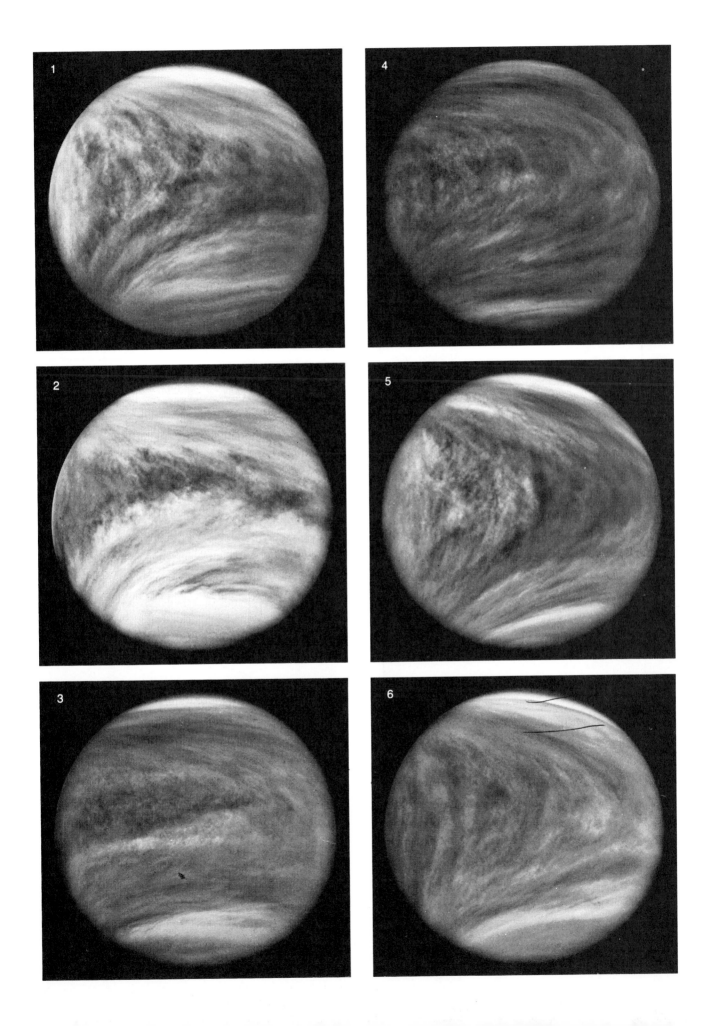

sphere of Venus is heated by the greenhouse effect, as has been proposed by Carl Sagan of Cornell University and James B. Pollack of the Ames Research Center of the National Aeronautics and Space Administration.

Above the troposphere lies a thinner upper atmosphere. In daylight the upper atmosphere is heated directly by ultraviolet radiation from the sun and the temperature therefore increases with height. Such an increase is also observed in the upper atmosphere of the earth, which is accordingly called the thermosphere. On the earth the thermosphere is present day and night; the large-scale rotation of the atmosphere with the planet carries the heated day-side upper atmosphere to the night side of the planet. On the night side of Venus, however, the thermosphere disappears; the upper at-

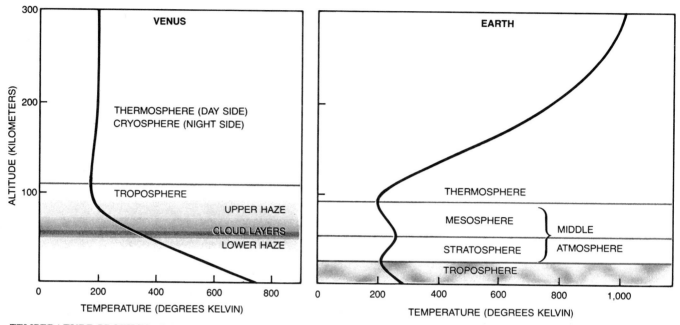

TEMPERATURE PROFILES of the atmosphere of Venus, earth and Mars are compared. The atmosphere of Venus (*left graph*) has two subdivisions. In the lower one, called the troposphere, the atmosphere is heated mostly by the greenhouse effect (the trapping of the heat the surface radiates). Hence the temperature decreases with altitude. In the upper subdivision, called the thermosphere, the atmo-sphere is heated directly by the absorption of solar radiation; the temperature increases with altitude. The profile for Venus is an average of daytime and nighttime measurements; actually the Venusian thermosphere could be said to disappear at night because the temperature of the atmosphere then decreases monotonically with altitude. The atmosphere of Mars (*right graph*) has the same two subdivisions

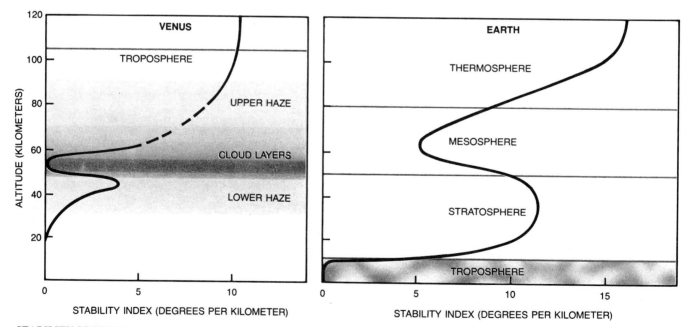

STABILITY PROFILES of the atmosphere of Venus, the earth and Mars indicate the degree to which the atmosphere at a given altitude resists convection. Basically a parcel of the atmosphere at any given altitude must support the weight of the atmosphere above it. Thus atmospheric pressure decreases with altitude. A parcel displaced upward will therefore expand, and as a result it will cool. If the gradient of adiabatic cooling (the cooling caused solely by the reduction in pressure) exceeds the atmosphere's temperature gradient, the displaced parcel will be cooler than its new surroundings and will sink back toward its original level. In that circumstance the atmosphere is stable against convection. If the adiabatic gradient is less than the temperature gradient, the displaced parcel will still be warmer than its surroundings. It will be buoyant and will continue to rise. The atmosphere is therefore unstable. The illustration plots the difference between the adiabatic gradient and the temperature gradient. Positive numbers indicate stability; zero suggests instability. (A negative value would not persist because convection would set in and tend to restore the balance.) On the earth the stability of the stratosphere confines

mosphere cools quickly after sunset to temperatures far below those of the troposphere. The nightly disappearance of the thermosphere of Venus was discovered on the Pioneer Venus mission. It has not yet been explained. The upper atmosphere of Venus rotates fast enough

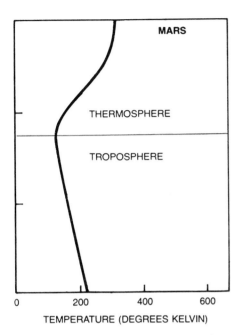

TEMPERATURE (DEGREES KELVIN)

as the atmosphere of Venus. The atmosphere of the earth (*middle graph*) is more complex than the others; it has a middle subdivision, in which the temperature profile attains a local maximum as a result of the absorption of ultraviolet radiation by the layer of ozone there.

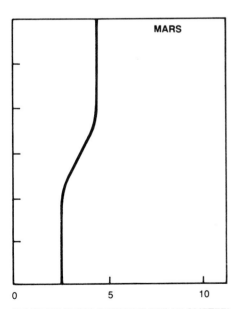

STABILITY INDEX (DEGREES PER KILOMETER)

convective mixing (and therefore clouds) to the troposphere. On Venus the pattern is similar, although a moderately stable layer lies subjacent to the clouds. On Mars the atmosphere is too thin for convective instability to develop. The data for the atmosphere of Venus and Mars were recorded by spacecraft. Data for the earth were provided by Richard Walterscheid of the Aerospace Corporation.

to carry a large amount of heat to the night side of the planet, but the heat is somehow lost. In any event we have suggested that the night-side upper atmosphere, which is the coldest part of Venus, be called the cryosphere.

The basic difference between the atmosphere of Venus and that of the earth is that the atmosphere of Venus is hot at the bottom and cold at the top, whereas on the earth the reverse is true. The lower atmosphere of the earth is not massive enough to sustain a large greenhouse effect. A second difference is that the earth's atmosphere has a middle region in which the temperature rises to a local maximum. The heating there results from the absorption of ultraviolet radiation by a layer rich in ozone.

In the troposphere of Venus horizontal variations in temperature are much smaller than the vertical variations. The greatest latitudinal differences the Pioneer probes detected were at the level of the upper clouds. Here the north probe, which entered the atmosphere at a latitude of 60 degrees, measured temperatures some 10 to 20 degrees lower than those measured by three other Pioneer Venus probes. The highest temperatures were measured by a probe that entered the atmosphere near the equator. The explanation of this pattern on Venus is the same as it is on the earth. In the equatorial region of each planet the sun's radiation enters the atmosphere at angles most nearly perpendicular to the surface. Hence the deposition of solar energy per unit area is greatest near the equator. On Venus the incident solar energy is absorbed mostly in the upper regions of the clouds.

Between an altitude of 70 kilometers and the beginning of the upper atmosphere at 100 kilometers the temperature gradient from the equator to the pole reverses: at these heights the polar regions are generally warmer than the equator. The reversal, discovered during the Pioneer Venus mission, has not been explained. A similar reversal of the temperature gradient is known to exist on the earth, where the winter pole is the warmest part of the middle atmosphere. There too the cause is not understood.

The variations of temperature with respect to longitude in the troposphere of Venus are much smaller than the variations with respect to latitude. This was shown by the measurements made by two other probes of the Pioneer Venus mission, the day and night probes, which entered the atmosphere at the same latitude (30 degrees south) but were separated by more than 110 degrees of longitude. The temperatures measured by these probes at any given altitude never differed by more than five degrees C. In the lowest 10 to 20 kilometers of the atmosphere the smallness of this difference results from thermal inertia; the lower atmosphere is so massive that it

retains most of its heat even during the long period it spends in darkness. At higher altitudes in the troposphere the rapid east-to-west circulation carries heat around the planet and reduces the temperature gradient.

Although the horizontal temperature differences are small, they must nonetheless be the source of the forces that drive atmospheric motions. In this respect the latitudinal gradient is particularly important. The solar energy entering the atmosphere of Venus warms the equator more than it warms the poles. On the other hand, the energy the planet emits in the infrared has virtually the same intensity at all latitudes because of the relatively small variation of temperature with respect to latitude. The imbalance between the incoming solar energy and the outgoing infrared energy would rapidly cool the poles and heat the equator were it not for large-scale motions of the atmosphere that carry heat from the equator toward the poles. Such motions are found on the earth, although the imbalance between the incoming and the outgoing radiation at the poles is less than it is on Venus.

How does heat move on Venus from the equator toward the poles? One would expect that a given parcel of the lower atmosphere in the equatorial latitudes of Venus would rise. Its place would be taken by adjacent parcels flowing in from the north and the south. In this way a steady current would ultimately become established, in which the atmosphere flows poleward at high altitude, bearing excess heat. After warming the polar regions the current would sink and then flow back toward the equator at a lower level.

North-south circulation of this kind is known as a Hadley cell. The evidence suggests that Hadley cells have formed on Venus at the level of the cloud deck. In particular, ultraviolet photographs made by the U.S. *Mariner 10* spacecraft when it flew by Venus in 1974 showed poleward motion at the tops of the clouds, whereas measurements made by all the Pioneer Venus probes showed winds toward the equator near the base of the clouds. A Hadley cell would lie at cloud heights on Venus because most of the incident solar energy is absorbed there. The cell would extend all the way from the equator to the pole because the slow rotation of the planet does little to deflect the circulation from its north-south trajectory. On the earth a Hadley cell lies just above the surface, the place where most of the solar energy is absorbed. Because the earth rotates rapidly the terrestrial Hadley cell extends no farther than the mid-latitudes, where the poleward transport of heat is taken over by complicated wavelike motions called baroclinic eddies.

A small fraction of the solar energy reaching Venus is absorbed at the surface, and since this energy is deposited predominantly at the equator, it too must be transported toward the poles. It is unlikely, however, that the cloud-level Hadley cell participates in this transport. The most important factor in analyzing the Hadley-cell circulation is the adiabatic gradient, which represents the cooling or heating a parcel of the atmosphere undergoes as a result of expansion or contraction when it is displaced upward or downward. If a parcel is displaced upward, for example, it enters a region where the pressure is lower, and it therefore expands and cools. In essence the heat in the parcel is distributed throughout a larger volume. The parcel may now be colder than the atmosphere around it. In that case it will be heavier than its surroundings and will tend to sink back to its original level. Under this condition the atmosphere is stable. On the other hand, the parcel may still be hotter (in spite of the adiabatic cooling) than the atmosphere around it. In that case it will continue to rise. The atmosphere is then unstable and is susceptible to overturning and the transport of heat by convection.

In evaluating the stability of an atmosphere one therefore compares two quantities: the rate at which temperature changes with altitude and the rate of adiabatic cooling. Wherever the latter is the smaller the atmosphere is unstable. When the atmosphere of Venus is analyzed in this way, a stable layer some 20 kilometers thick is found to lie immediately below the clouds. Since vertical motions would be suppressed in this layer, it is doubtful that a single Hadley cell could extend from the surface to the clouds.

A separate Hadley cell might operate near the surface, but the presence of such a cell is still a matter for speculation. The north-south winds measured deep in the atmosphere by the Pioneer Venus probes are small; they have magnitudes of only a few meters per second and their directions reveal no large-scale pattern. If a Hadley cell does lie deep in the atmosphere, the mean north-south circulation in the atmosphere of Venus would have to consist of at least three Hadley cells stacked one on top of the other because the cloud-level cell and the deep cell would drive at least one counterrotating cell between them. If there is no deep Hadley cell, the solar energy that penetrates to the surface must be carried poleward by the net effect of eddy motions.

Since Venus rotates slowly, one might expect north-south circulation to constitute the totality of the atmospheric circulation. For one thing, seasonal differences on the planet should be insignificant. Venus' elliptical orbit deviates only slightly from a circle, so that the planet maintains an almost constant distance from the sun. Moreover, the planet's axis of spin is almost perpendicular to the plane of the orbit. At any latitude, therefore, the deposition of sunlight changes only slightly through the year.

Surprisingly, the expectation of a simple pattern of circulation in the atmosphere has been contradicted by the evidence. The mean north-south circulation is overshadowed by the far stronger east-to-west wind, which begins some 10 kilometers above the surface and extends to heights of 90 kilometers or more. Up to the height of the clouds the wind speed increases with altitude. It reaches a velocity of 30 meters per second at an altitude of 30 kilometers and 100 meters per second at cloud heights of 60 kilometers. At the latter speed the wind gives rise to a four-day superrotation of the ultraviolet markings in the clouds. At cloud heights, moreover, the superposition of the westward circulation and the weaker Hadley circulation gives rise to a vortex that converges at each pole. The vortex was first recognized in a composite photograph of the south pole that Verner E. Suomi and Sanjay S. Limaye of the University of Wisconsin at Madison constructed in 1978 from ultraviolet photographs of Venus made by *Mariner 10* at latitudes nearer the equator.

The superrotation had first become apparent when observations from the earth showed that large-scale features in the atmosphere of Venus circle the planet about once every four days. The features could be seen in ultraviolet photographs of the planet. Until the Venera and the Pioneer Venus probes measured the wind speed, however, one could not be certain whether the movement of the ultraviolet features represented bulk motion of the atmosphere or was instead the propagation of a wave—a sloshing, so to speak—in an otherwise placid atmosphere. It is now generally accepted that the large-scale features are caused by planetary waves but that the sloshings propagate slowly with respect to the rotation of the atmosphere. What propels the features around the planet is indeed the westward wind.

The precise nature of the planetary waves remains hard to determine. The problem is that the substances responsible for absorbing ultraviolet radiation and thereby making the large-scale fea-

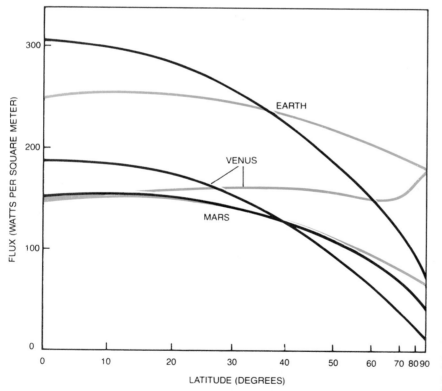

RADIATION BALANCE between incoming and outgoing solar energy is attained on Venus, the earth and Mars because the atmosphere of each planet transports heat from its equator to its poles. At the equator the amount of radiation each planet receives from the sun (*black lines*) exceeds the amount it emits in the form of infrared radiation (*colored lines*). The excess is carried toward the poles. At the poles the amount of radiation the planet emits would exceed the amount it receives if it were not for the poleward flow. Because the radiation balance has not been measured extensively on Mars the curves for that planet are hypothetical. The difference between the curve representing incoming radiation and the curve representing outgoing radiation on Mars is small because the atmosphere of Mars is too tenuous to hold much heat.

tures visible have not all been identified. Specifically, the absorption of radiation at short ultraviolet wavelengths can be attributed to sulfur dioxide gas, but the absorption at longer wavelengths remains to be explained. The absorber at longer wavelengths cannot be sulfuric acid, the main constituent of the clouds, because sulfuric acid is transparent to ultraviolet radiation. The absorbers of the radiation may turn out to be concentrated below the cloud tops in areas of the planet that are bright in the ultraviolet. Vertical motions induced by the passage of a planetary wave could then give rise to contrast in the ultraviolet image of the planet by lofting the absorbers to a higher altitude, where their absorption of ultraviolet radiation would darken parts of the image.

The extent to which the four-day circulation persists with increasing height above the cloud tops is also not well known. The Pioneer Venus data suggest that the wind speed reaches a maximum of about 150 meters per second at an altitude of 70 kilometers and then decreases with height between 70 and 90 kilometers.

Above 150 kilometers the circulation of the atmosphere has been measured only by indirect methods. For example, a mass spectrometer on board the Pioneer Venus orbiter measures the density of substances such as nitrogen, carbon dioxide and helium high in the atmosphere. In addition the orbiter is tracked as it revolves about the planet. The slow decay of its orbit is caused by drag, which in turn depends on the density of

PATTERN OF WINDS in the lower atmosphere of Venus differs markedly from the pattern on the earth. On Venus (*upper drawing*) the data recorded by probes suggest that at cloud heights of roughly 60 kilometers the atmosphere circulates in a current that rises at the equator, travels poleward, sinks near the pole and returns to the equator. The pattern, which is called a Hadley cell, represents the poleward flow of the heat the sun deposits in the clouds. Two more Hadley cells are shown, but their presence is hypothetical. The cell nearest the surface may be needed to carry heat in the lower atmosphere; the stable layer just below the clouds seems to prevent the heat deposited near the surface from rising to cloud heights. If the bottom cell exists, there must be a middle cell in counterrotation between the other two. The north-south flow in the Hadley cells has a speed of only a few meters per second. Superimposed on this motion is the westward rotation of the atmosphere (*white lines*), which attains a speed of approximately 100 meters per second at the altitude of the clouds. On the earth (*lower drawing*) a Hadley cell lies near the surface. It extends only to mid-latitudes, where the rotation of the earth disrupts its north-south path. From there the poleward transport of heat is the net effect of ever changing patterns of eddies.

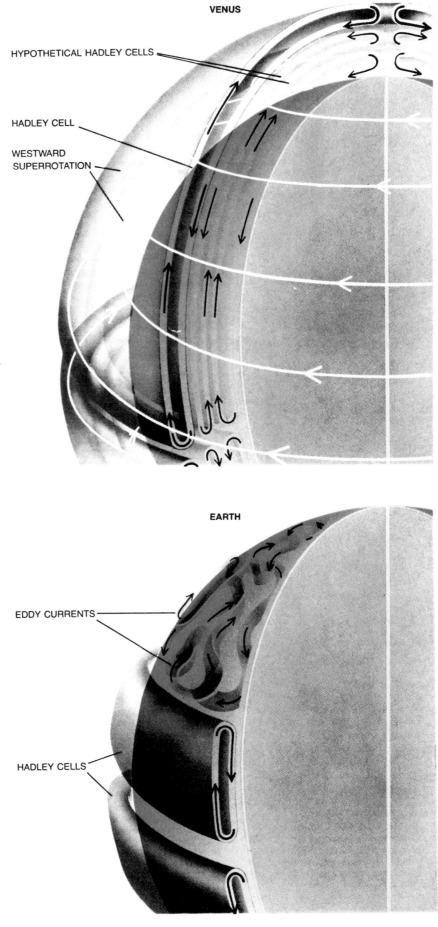

VENUS

HYPOTHETICAL HADLEY CELLS

HADLEY CELL

WESTWARD SUPERROTATION

EARTH

EDDY CURRENTS

HADLEY CELLS

the atmosphere. By both of these methods it has been shown that the place of lowest density in the upper atmosphere lies well to the west of the point opposite the sun on the night side of the planet.

This finding suggests that an earlier hypothesis should be amended. Before the Pioneer Venus mission many investigators had expected that the main influence on the circulation of the upper atmosphere of Venus would turn out to be the heating of the part of the upper atmosphere facing the sun. On that hypothesis a parcel of the upper atmosphere would ascend in the region facing the sun, flow to the night side of the planet and descend. The overall pattern of circulation would thus be symmetrical about a line connecting the center of the sun with the center of Venus. The Pioneer Venus measurements imply a more complex pattern, in which a westward superrotation may deflect the descending limb of the circulation.

Why is the dominant circulation at all altitudes a westward superrotation? That remains the greatest mystery about

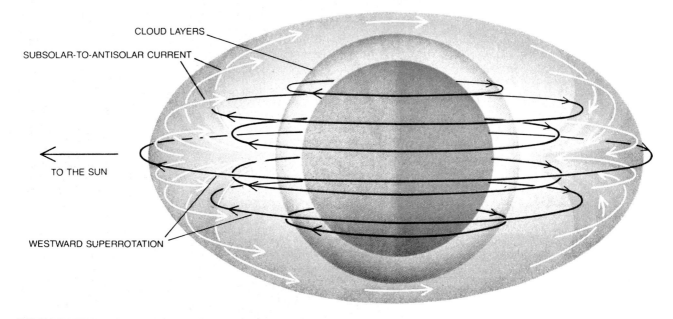

WINDS IN THE UPPER ATMOSPHERE of Venus are hypothesized to result from the superposition of two basic flows: the westward rotation of the atmosphere (*black lines*) and a symmetrical circulation (*white arrows*) in which the winds ascend on the day side of the planet and descend on the night side. Although the winds in the upper atmosphere have not been measured directly, the superposition shown is supported by measurements indicating that the density of the upper atmosphere is lowest at a point on the night side of the planet well to the west of the point opposite the sun. It is as if the sinking current were being deflected by the westward current.

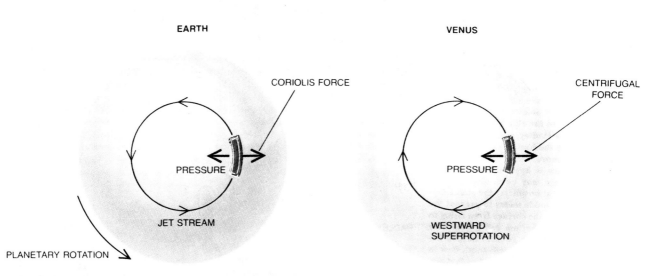

BALANCE OF FORCES that maintains atmospheric motions on a rapidly rotating planet (such as the earth, Mars, Jupiter or Saturn) differs from the balance on a slowly rotating planet such as Venus. In each case the greater deposition of heat at the equator of the planet establishes a pressure gradient that drives the atmosphere toward the pole. On the earth (*left*) the pressure is opposed primarily by the Coriolis force: the deflection of a parcel of the atmosphere at a right angle to its trajectory because it is on a rotating sphere. The earth's jet stream, for example, is supported by such a balance; the Coriolis force drives the jet stream toward the Equator and counteracts the pressure that drives it toward the pole. The condition is called geostrophic balance. On Venus (*right*) the Coriolis force is negligible. Instead the motion of a parcel circling the planet as part of the westward rotation of the atmosphere subjects it to a centrifugal force, which balances the poleward pressure by pushing the parcel away from the planet's spin axis. The condition is cyclostrophic balance.

the atmosphere of Venus. The essential difficulty is that one expects an atmosphere to rotate more or less in step with the solid planet below it; if it rotates faster, one must imagine a mechanism by which angular momentum is transported upward from the surface of the planet. In effect the surface of the planet must push the atmosphere.

A mechanism that might account for the upward transfer of angular momentum is the moving-flame effect. Imagine a flame under a vessel holding a layer of fluid, and imagine initially that the flame stands still. Over a period of time a convection cell develops, in which the fluid directly over the flame rises and the fluid around this upwelling sinks. Since the flame is stationary, both the ascending and the descending limbs of the convection cell are vertical. Now imagine that the flame is moved continuously from left to right under the layer. The bottom of the cell will come to lie to the right of the top. Hence the limbs of the cell will be tilted: the upwelling fluid will move to the left as it rises and the descending fluid will move to the right as it sinks. In laboratory experiments employing a ring of fluid it has emerged that these effects reinforce each other, and the entire fluid flows to the left. The fluid flows, then, in a direction opposite to the direction in which the flame is moved.

On Venus, which rotates toward the west, the sun moves toward the east in the sky. A moving-flame effect caused by this eastward motion of the sun might therefore build up the westward superrotation. To be sure, the sun is above the atmosphere of Venus whereas the flame is under the fluid layer. On Venus, however, the situation is complicated by the stable layers of the atmosphere, which tend to resist the formation of convection cells. An example is the layer below the clouds. In a calculation one of us (Schubert) made with Richard E. Young of the Ames Research Center, the hypothetical convection cells on Venus turn out to be not merely tilted: their vertical profile has the shape of a boomerang. According to the calculation, it is conceivable that such cells could impart to the atmosphere a net westward rotation.

Another hypothesis, suggested by Peter J. Gierasch of Cornell University, is that the Hadley circulation transports angular momentum high into the atmosphere. In this view a parcel of the atmosphere that rises at the equator takes with it an angular momentum corresponding to that of the surface at the equator, and as the parcel moves poleward in a Hadley cell it gives up some of its momentum to the surrounding atmosphere. When the parcel returns to the surface, it regains momentum, and so the cycle continues. Perhaps the cycle works in spite of a stacking of three or

SMALL ATMOSPHERIC WAVES are visible in ultraviolet photographs made by the Pioneer Venus orbiter. Here the waves are a file of dark diagonal lines at the center of the image. Each line is roughly 1,000 kilometers long and 200 kilometers from its neighbor on each side.

more Hadley cells. An analysis of ultraviolet photographs made during the *Mariner 10* flyby of Venus shows that at latitudes of from 40 to 45 degrees the cloud markings were moving westward in 1974 somewhat faster than they would move if the atmosphere were rotating rigidly, that is, like a solid body. It has been suggested by William B. Rossow and his colleagues at the Goddard Institute for Space Studies that angular momentum carried upward by the Hadley circulation might have built up this jet and that the turbulence around the jet might disperse the angular momentum throughout the atmosphere. In the Pioneer Venus data, however, no such jet has been found.

One thing is certain: the atmosphere of Venus responds to a balance of forces fundamentally different from that of the atmosphere of the earth and Mars, or for that matter of Jupiter and Saturn. The reason is the slow rotation of Venus. On any planet the poleward

transport of heat by winds corresponds to a gradient in atmospheric pressure: the pressure is greater at the equator than it is at the poles. On a rapidly rotating planet such as the earth the resulting poleward drive on a given parcel of the atmosphere is balanced by the Coriolis force: the deflection of a parcel of the atmosphere at a right angle to its trajectory because it is on a rotating sphere. (To an observer some distance from the planet it would be apparent that the deflection corresponds to the rotation of the planet's coordinate grid of latitude and longitude below the wind. To an observer in motion with the planet it is the wind that seems to be deflected.) The Coriolis force on a given parcel is proportional to the velocity of the parcel.

Since Venus rotates slowly, the Coriolis force is negligible. Instead, as was first noted by Conway B. Leovy of the University of Washington, the poleward drive on a parcel of wind that circles the planet from east to west as part of the global superrotation of the atmosphere is balanced by the centrifugal force that tends to push the parcel away from the spin axis of the planet, and therefore toward the equator. Because the centrifugal force depends on the square of the velocity of the parcel, it is fundamentally different from the Coriolis force.

The ultraviolet photographs of the atmosphere of Venus reveal a variety of atmospheric markings, which presumably will find explanation in accord with the unusual balance of forces there. The largest of the markings are the planet-wide features that first revealed the four-day westward circulation. We noted above the association of such features with planetary waves. Doubtless the waves propagate vertically as well as horizontally; they can therefore carry momentum upward in the atmosphere,

and so they may have a role in fostering the four-day circulation. On a smaller scale some of the Pioneer Venus photographs have shown trains of waves. In one case the waves consisted of dark bands about 1,000 kilometers long separated from one another by about 200 kilometers. The waves were found in the southern mid-latitudes of Venus and were arrayed at a large angle to the planet's parallels of latitude.

Polygonal features are apparent in some of the ultraviolet photographs. They are undoubtedly convection cells. Until it is possible to identify all the atmospheric absorbers of ultraviolet radiation that modulate the ultraviolet images, it cannot be said whether the dark core of each such cell is the place where the atmosphere is upwelling and the

bright rim of the cell is the place where the atmosphere sinks. The reverse might be the case. In any event the convection cells are seen mostly in the part of the atmosphere that directly faces the sun or else to the west (that is, downwind) of that region. The absorption of solar radiation there is apparently sufficient to induce convection near the tops of the clouds. An analysis of the stability of the atmosphere based on measurements made by the Pioneer Venus probes that entered the atmosphere in the early morning shows that a layer about 10 kilometers below the cloud tops is unstable and capable of convective overturning. This unstable layer must grow thicker throughout the morning, so that by afternoon the tops of the convective cells are visible from above the clouds.

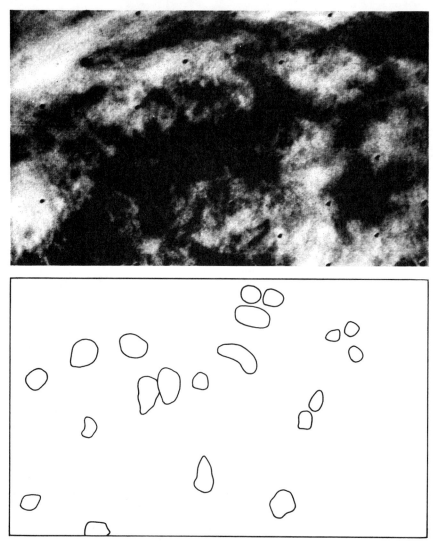

BRIGHT-RIMMED CELLS on the day side of Venus are visible in an ultraviolet photograph made by the *Mariner 10* spacecraft when it flew by the planet in 1974. Each cell is a few hundred kilometers across, and some cells are distinctly polygonal. They are thought to be convection cells: places where the atmosphere overturns. The overturning is presumably driven by solar energy absorbed near the top of the clouds. Further knowledge awaits the identification of all the trace components of the atmosphere whose absorption of radiation gives rise to the ultraviolet markings. The photograph was provided by Michael J. S. Belton of the Kitt Peak National Observatory. The accompanying map shows the positions of prominent convection cells.

Waves and convection cells are two examples of eddy motions: short-term variations superimposed on the mean state of the atmosphere. Such motions are present at all spatial scales and at every altitude. Evidence of them is found not only in the ultraviolet photographs but also in the records of temperature and wind velocity made by the Pioneer Venus probes. The atmosphere of Venus is also surprisingly variable on time scales as long as months or years. An example of the long-term variability is an apparent change in the dominant westward circulation. Ever since the Pioneer Venus mission entered the atmosphere of Venus in December, 1978, the atmosphere at cloud level has been rotating like a rigid body, but at the time of *Mariner 10* some four years earlier the mid-latitudes were rotating faster than the rest. (This constituted the jet mentioned above.) It is difficult to understand what might be responsible for such a change when one considers the steadiness of the influx of solar radiation on Venus and the gentle undulation that is now known to characterize the surface of the planet.

The exploration of the atmosphere of Venus has already revealed its high temperature near the surface, its westward superrotation, its cloud-level Hadley cell, its polar vortexes, its night-side cryosphere and its warm poles above the clouds. A collaboration of Russian and French groups is planning to place a balloon well below the clouds of Venus in 1984. The balloon will be tracked as the east-to-west wind sweeps it across the daylight side of the planet. A mission planned by the U.S. for the mid-1980's may carry instruments to gather data on the upper atmosphere, which the Pioneer Venus orbiter is also continuing to probe. The exploration of the atmosphere of Venus has already been rewarding beyond expectation, and the findings that remain unexplained surely warrant further effort.

3

The Surface of Venus

by Gordon H. Pettengill, Donald B. Campbell
and Harold Masursky
August, 1980

*Shrouded by clouds, it has now been mapped by radar
from the earth and from a spacecraft in orbit around
Venus. The images suggest a geology intermediate
between that of the earth and that of Mars*

Over the past 15 years the surface of Venus, perpetually hidden by its deep cover of clouds, has been intensively probed by radar signals beamed from the earth, and over the past year and a half it has also been probed by signals from a spacecraft placed in orbit around it. The returning radar reflections reveal that the planet's surface is marked by geologic features suggestive of impact cratering, volcanism and tectonic activity. Because of orbital and rotational synchronicities and other constraints the radar view of Venus from the earth is effectively limited to less than one complete hemisphere of the planet. On the other hand, the Venus orbiter has covered the complete circumference of the planet between 74 degrees north latitude and 63 degrees south latitude. Within that broad band, encompassing 93 percent of the planet's surface, the Venus orbiter can measure altitudes to an accuracy of within 200 meters. The spacecraft has found that Venus is remarkably flat: 60 percent of its surface falls within a height interval of only one kilometer. One feature, however, rises 11 kilometers above the surrounding plain and therefore is taller than Mount Everest. Named Maxwell, it is possibly a shield volcano.

When early telescopes showed that Venus was not only bright but also apparently shrouded in clouds, man's imagination was quick to populate the planet's surface with dinosaurlike creatures living in steamy swamps. The planet is only 72 percent as far from the sun as the earth is, so that one could expect Venus to be substantially hotter than the earth and the clouds to be condensed water vapor. Later, as spectrographs and other instruments showed that the atmosphere of Venus consists primarily of carbon dioxide and is almost devoid of water vapor, most of the early fancies died. One of them, however, was borne out. Measurements of the planet's thermal radiation, made at radio wavelengths long enough to penetrate the clouds, have shown that its sur-

face is indeed hot: 475 degrees Celsius (about 900 degrees Fahrenheit).

By 1960 radar systems had been constructed with sufficient power to detect radio echoes from Venus on the earth. Operating at wavelengths of between 12 and 70 centimeters, the radar systems penetrated the planet's atmosphere and proved that a true surface exists. Although the early systems lacked the resolution to determine the radius of the solid surface with any precision, they did provide the first clues to two general properties of the surface, namely its mean reflectivity and its average "roughness" at radar wavelengths. The observed reflectivity of about 15 percent is typical of many rocks on the earth.

Working with radar observations made in 1961, William B. Smith of the Lincoln Laboratory of the Massachusetts Institute of Technology looked for evidence of frequency broadening in the returning echoes. If Venus is rotating, a differential Doppler shift should be imparted to components of the radar signal that are reflected from different areas of the surface. Signal components that strike an approaching surface will be increased in frequency and those that strike a receding surface will be decreased. The overall effect is to broaden the frequency spectrum of the echoes. Smith concluded that the frequency broadening of the echoes was small

and that possibly it decreased as Venus passed between the earth and the sun.

At that time essentially nothing was known about the rate at which Venus rotates. On the basis of his observations Smith suggested that the rotation might be retrograde: in a direction opposite to that of the planet's motion around the sun. A year and a half later, in 1962, Roland L. Carpenter and Richard M. Goldstein of the Jet Propulsion Laboratory of the California Institute of Technology showed that Venus' rate of rotation is very low, having a period equal to about 240 earth days, and that the rotation is indeed retrograde.

The next major advance in the radar study of Venus came with the completion of the Arecibo radar antenna in Puerto Rico (with a diameter of 300 meters and operating at a wavelength of 70 centimeters), the deep-space tracking antenna at Goldstone, Calif. (with a diameter of 64 meters and operating at 12.5 centimeters) and the Haystack radar system in Massachusetts (with a diameter of 43 meters and operating at 3.8 centimeters). By 1966 the accuracy of the radar ranging of Venus with these instruments had improved to better than within one kilometer, or nearly one part in 10^8 of the distance to the planet. Measurements at this accuracy were carried out over several years at Arecibo by

RADAM IMAGE OF VENUS was assembled from observations made at the Arecibo Observatory in Puerto Rico in 1975 and 1977, when Venus made close approaches to the earth. At such times Venus always presents nearly the same "face" to the earth, so that only slightly less than one hemisphere, the region centered on 320 degrees east longitude, can be successfully mapped. The empty band below the equator is a region where in 1975 and 1977 the imaging method was unable to resolve the radar echoes. Echoes from that region were successfully resolved in observations made during the close approach of this June, but the data are still being processed. The observing geometry in 1975 made the northern hemisphere of the planet slightly more accessible than the southern, as is evident in this Mercator projection. The image brightness is proportional to the degree of surface roughness with dimensions of a few centimeters, except at low latitudes, where meter-scale slopes also influence the reflected signal. The lateral resolution in this mosaic varies between 10 and 20 kilometers. (A degree of latitude on Venus equals 106 kilometers.) The bright area at the upper right, named Maxwell (centered at 65 degrees north latitude, five degrees east longitude), and the area at the left center, named Beta (25 degrees north, 283 degrees east), have nearly saturated the display in this representation. The two areas are shown separately at reduced contrast in the illustrations on page 34.

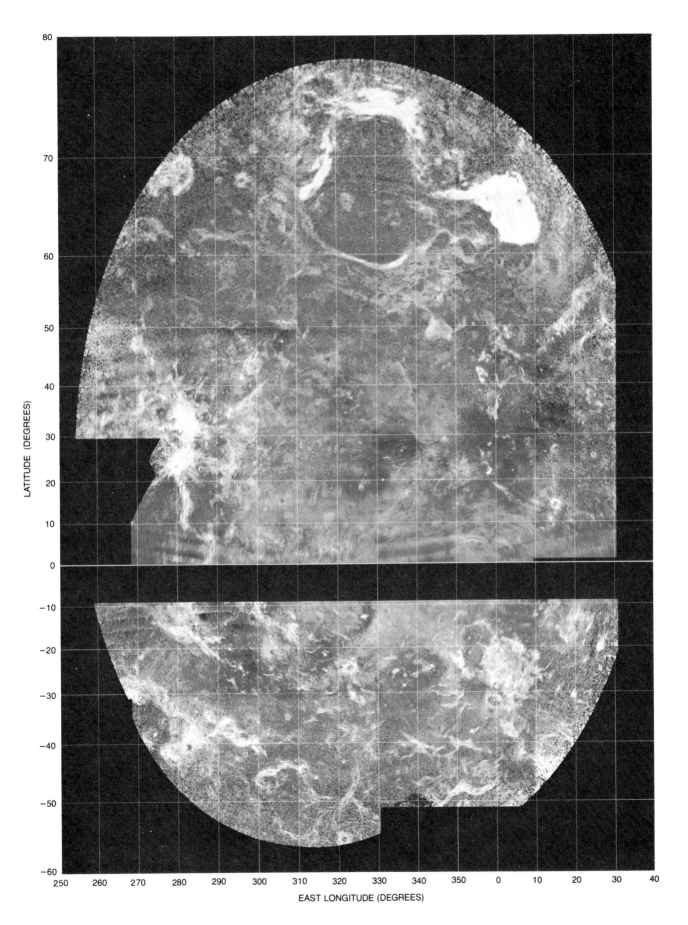

EAST LONGITUDE (DEGREES)

LATITUDE (DEGREES)

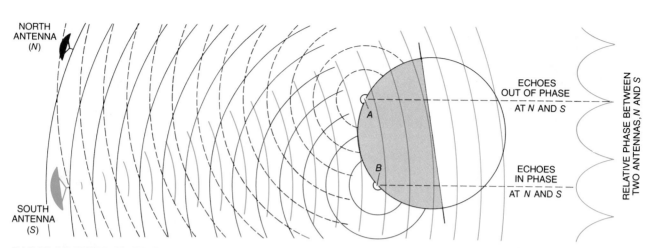

RADAR MAPPING OF VENUS by the authors and their associates exploits interferometry to distinguish between echoes arising from two separated locations on the surface of the planet. The transmitter in these experiments is at the 300-meter radio telescope at Arecibo. The emitted radio signal has a nominal wavelength of 12 centimeters and a power of 400 kilowatts. The outgoing radio beam (*color*) consists of a continuous-wave signal about five minutes in duration that has been systematically varied in phase. When the radar echoes return from Venus after a round trip of roughly five minutes, the precise travel time can be deduced by searching for the phase patterns that were embedded in the original outgoing signal. The echo is

recorded simultaneously at the 300-meter antenna (*S*) and at a 30-meter "passive" antenna (*N*) 10 kilometers to the north. A point on Venus such as *B*, for which the round-trip distances *SBS* and *SBN* differ by an exact multiple of the wavelength, gives the same received phase at the two antennas. Another point, *A*, for which the corresponding paths *SAS* and *SAN* differ by an odd multiple of half wavelengths, gives echoes that are out of phase. (Here the wavelength is much exaggerated.) When the echoes received at the two antennas are combined coherently with appropriate adjustments in their relative phases, one can suppress the echo from either *A* or *B*. Here the signals from point *A* "cancel," leaving only the signals from *B* visible.

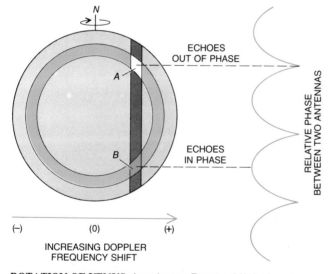

ROTATION OF VENUS gives rise to a Doppler shift in the returning signals. Echoes from approaching sites are shifted to a slightly higher frequency; echoes from receding sites are shifted to a slightly lower one. Two sites, *A* and *B*, produce echoes with the same time

delay and Doppler shift. The strength of the scattering from a conjugate pair of locations such as *A* and *B* is established by analyzing the radar echo to extract those components with the same time delay and Doppler shift. When *A* is in an interferometer null, only *B* is visible.

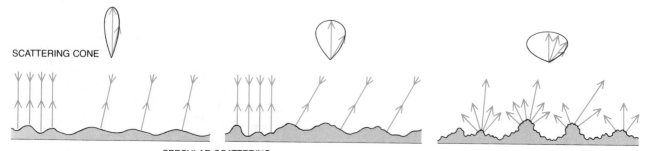

CHARACTER OF A SURFACE can be inferred from the distribution of the scattered power at varying angles. An undulating surface, smooth at the scale of the observing wavelength, scatters back radar energy only where a facet lies at right angles to the observer. For gen-

tle slopes this produces a narrow scattering cone (*left*). Where a surface has somewhat steeper facets the cone is wider (*middle*). A surface that is rough at the scale of the wavelength scatters diffusely with an intensity largely independent of the angle of observation (*right*).

Rolf B. Dyce and two of us (Campbell and Pettengill) and at Haystack by Richard P. Ingalls and one of us (Pettengill). In collaboration with Irwin I. Shapiro and other workers at M.I.T. we deduced that the mean surface radius of Venus is close to 6,050 kilometers. As the orbits of the earth and Venus came to be better known from the fitting of the motion of the two planets to long sequences of radar data it was eventually possible to observe variations in the radius of Venus as the reflecting region migrated around the equator of the planet.

An unexpected dividend of the radius measurement was the contribution it made to determining the atmospheric pressure at the surface of Venus. The pressure determination is a classic example of the power of combining apparently unrelated lines of inquiry. One of the tasks of the 1967 mission of *Mariner 5* was to observe the refraction of the spacecraft's radio signal by the atmosphere of Venus as the craft passed behind the planet. From the refraction data Arvydas J. Kliore and his colleagues at the Jet Propulsion Laboratory calculated the atmospheric pressure and temperature as a function of distance from the planet's center of mass. Because the dense lower atmosphere of Venus is both highly refractive and absorbing, however, Kliore was unable to extend the spacecraft measurements below altitudes where the pressure is more than a few times the atmospheric pressure at the surface of the earth.

An extrapolation to the surface of Venus was made possible by taking account of the surface temperature (from the radio-emission measurements) and the precise location of the surface (from our radar-ranging measurements). The major uncertainty in the extrapolation was ignorance of the composition of the atmosphere near the surface. On the assumption that the composition is dominated by carbon dioxide, Kliore and his colleagues calculated that the surface pressure on Venus is almost 100 times the surface pressure on the earth. This prediction was subsequently verified directly by Russian and (much later) U.S. probes that descended through the atmosphere and reached the surface.

Hot, dry and under tremendous atmospheric pressure, the surface of Venus is unique in the solar system. How could one examine in detail this remote, beclouded and inhospitable area nearly the size of the earth? Clearly the most feasible approach was again to use radar. The challenge was to extend and increase the sensitivity of the techniques that had successfully defined the planet's radius and rate of rotation.

Since points on the surface of a rigid body move in a highly predictable way as the body rotates, one can correlate with a specific surface location the time it takes a component of the radar signal

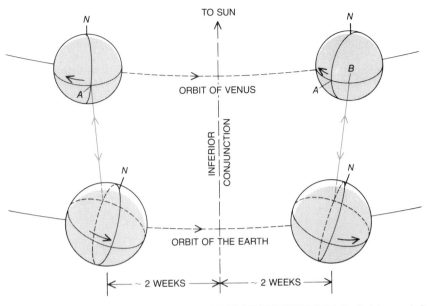

RADAR OBSERVATION OF VENUS WITH HIGH RESOLUTION is limited to a period of about four weeks centered on the time of inferior conjunction, when Venus is on a line between the sun and the earth and therefore at its closest to the earth. A minimum distance to the target is necessary because the intensity of radar echoes falls off inversely as the fourth power of the target's distance. An inferior conjunction comes every 19 months. Since Venus makes one retrograde rotation every 243 earth days, it rotates from east to west (with respect to the stars) about 42 degrees, or 1.5 degrees per day, during the prime four-week observing period. For an observer on the earth, however, about a third of this rotation is canceled because Venus is moving from east to west (with respect to the stars). Thus two weeks after inferior conjunction a point on the surface of Venus, A as viewed from the earth, will have rotated only about 27 degrees eastward from B, its position two weeks before inferior conjunction.

to make its round trip and the Doppler shift in its frequency at reflection. With computer analysis of the data one can map variations in the scattering of the radar signal over a large fraction of the planetary surface provided only that the radar system has sufficient sensitivity at the desired resolution.

In our studies the outgoing beam is transmitted from the 300-meter Arecibo antenna and consists of a continuous-wave signal lasting for about five minutes that incorporates carefully orchestrated rapid reversals in phase. By searching the echo for a corresponding pattern of phase reversals we can precisely measure the round-trip delay of the signal. When the transmitted beam reaches Venus, it has spread out to about twice the diameter of the planet. The radar echoes are recorded at the Arecibo antenna, whose transmitter is turned off while the signal is being received, and simultaneously at a smaller dish, 30 meters in diameter, that is 10 kilometers to the north. Doppler shifts in the recorded signals are established by frequency analysis. Differences in phase between corresponding components of the echo received at the two sites are then analyzed by computer to help sort out and map the locations of the reflecting surfaces on Venus.

What properties affect the scattering of radio waves? If a surface is smooth at the scale of the observing wavelength, the reflection is specular, or mirrorlike. The radar beam can be visualized as

a searchlight. If the radar reflection is viewed by the same antenna that launches the radar pulse, energy that is reflected specularly will be recorded only where smooth facets of the surface lie at right angles to the axis of the antenna. At radar wavelengths the moon, Mercury, Venus and Mars appear to be densely covered by facets that vary in dimension from a few wavelengths to hundreds or even thousands of wavelengths and that are tilted at random to the local horizontal. The same would be true of many points on the earth that are not covered by water or vegetation.

Radar echoes from the inner planets are therefore dominated by specular reflection when the backscattering angle of incidence is approximately perpendicular to the mean surface. The phenomenon can be likened to sunlight glinting from the sides of wavelets on the ruffled surface of a body of water. The rate of decrease in the power of the backscattered signal as the viewing angle departs from the perpendicular depends on the steepness of the slopes of the facets. The total power of the scattered signal, on the other hand, depends primarily on the reflectivity of the surface.

Surface irregularities that have sharp edges or that are small compared with the incident wavelength give rise to diffuse or incoherent scattering that spreads the incident power thinly over a wide range of emerging angles. When the angles of incidence are more than about 20 degrees from the mean local vertical, scattering from small irregularities usually dominates over specular reflection. Accordingly observations of radar backscattering at typical decimeter wavelengths made at viewing angles within 10 degrees of the local vertical are sensitive primarily to surface undulations with dimensions larger than a meter, and observations made at more oblique viewing angles are sensitive primarily to surface structures with dimensions of about a centimeter. By and large the variations in radar scattering seem to be dominated by the geometry of the surface rather than by changes in its reflectivity, which is a function of the dielectric discontinuity between the planet's surface and its atmosphere.

Radar images of Venus made from the earth are assembled from observa-

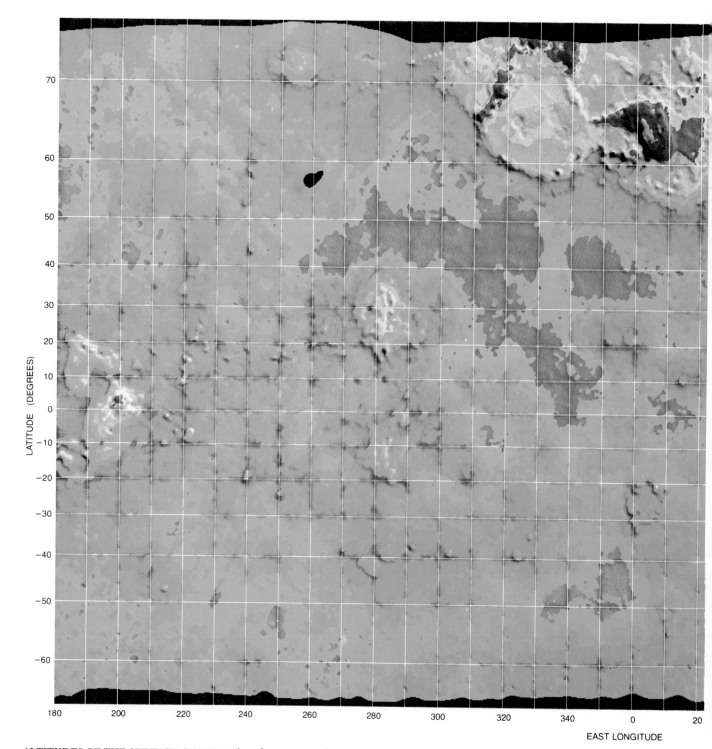

ALTITUDES OF THE SURFACE OF VENUS have been measured by *Pioneer Venus I*, a small spacecraft that was placed in orbit around the planet on December 4, 1978. A radar altimeter aboard the orbiter measures the height above the surface of Venus whenever the craft passes below an altitude of 4,700 kilometers on an orbit that carries it from about 150 kilometers at periapsis to about 66,000 kilometers at apoapsis. The orbit limits coverage to the 93 percent of the planet's surface that lies between the latitudes 74 degrees north and 63 degrees south. The raised appearance of this computer rendering is created by modulating the intensity of the color contours, as though the sun were shining on the surface from the upper right, and it serves to emphasize the steeper slopes. Average radius of Venus is 6,051.4 kilometers, coded a blue-green in the altitude color scale. The deepest violet corresponds to a planetary radius of less than 6,049.5 kilome-

tions taken over a wide range of scattering angles. The rotational axis of Venus is always nearly at right angles to the line of sight from the earth, so that scattering data from high latitudes on Venus are obtained at large angles of incidence; therefore the variations in surface structure the data disclose have dimensions of less than a wavelength. At lower latitudes, where the radar-viewing angle is more nearly straight down, the scattering is modulated by surface undulations and inclinations that have dimensions of more than a wavelength. With decreasing distance to the poles of the planet coverage is limited by the increasing depth of atmosphere to be penetrated and the resulting high absorption of the radar signal. The backscattering efficiency also drops for radio waves that strike the surface at shallow angles,

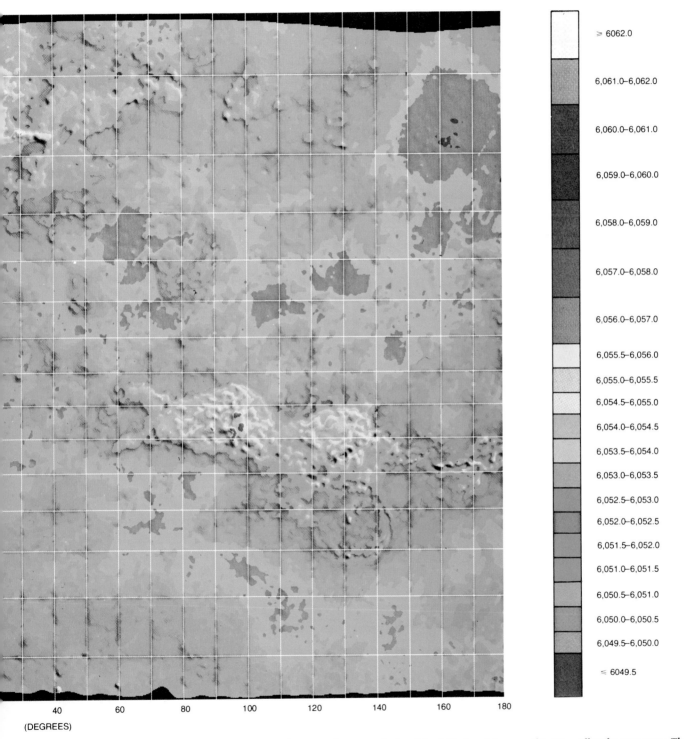

	≥ 6062.0
	6,061.0–6,062.0
	6,060.0–6,061.0
	6,059.0–6,060.0
	6,058.0–6,059.0
	6,057.0–6,058.0
	6,056.0–6,057.0
	6,055.5–6,056.0
	6,055.0–6,055.5
	6,054.5–6,055.0
	6,054.0–6,054.5
	6,053.5–6,054.0
	6,053.0–6,053.5
	6,052.5–6,053.0
	6,052.0–6,052.5
	6,051.5–6,052.0
	6,051.0–6,051.5
	6,050.5–6,051.0
	6,050.0–6,050.5
	6,049.5–6,050.0
	≤ 6049.5

40   60   80   100   120   140   160   180

(DEGREES)

ters, which embraces much less than 1 percent of the mapped area. The color changes at each half-kilometer interval until brown is reached at an altitude of 6,056 kilometers, beyond which the interval increases to one kilometer. The accuracy of the measurements is within .2 kilometer. The small black area at the upper left should be filled in as the *Pioneer Venus I* mission continues. The highest elevation so far recorded by the spacecraft lies just above 6,062 kilometers in the white dot in the pink area within Maxwell at the top center. The mountain's 11-kilometer height above the average radius of the planet exceeds by two kilometers the height of Mount Everest above sea level. The region between longitudes 260 degrees and 30 degrees east should be compared with the radar image of the region on page 27. This map and others presenting the Venus-orbiter radar data were prepared by Eric Eliason of the U.S. Geological Survey.

which further reduces the strength of echoes from high latitudes.

Since Venus presents different faces to the earth as it rotates, one might think that in due course its entire surface could be mapped. Such is not the case. The strength of the transmitted radar energy falling on the target and the proportion of that energy received as an echo back on the earth both fall off with the inverse square of the distance to the target, so that the strength of the echo is related to the strength of the original transmission inversely as the fourth power of the distance. High-resolution radar mapping is therefore limited to a few months at inferior conjunction: the time, coming about once every 19 months, when Venus is on a line be-

tween the earth and the sun and is therefore closest to the earth. At such a time the combined effect of Venus' retrograde rotation and the planet's westward motion through the sky causes the subearth point on Venus, defined by a line passing through the center of both planets, to change by less than .8 degree per day. Hence in the most favorable observing period there is little apparent rotation. Even so, one would expect the subearth point to migrate systematically from one inferior conjunction to the next. Curiously, this does not happen. Venus rotates almost exactly four times, as it is viewed from the earth, between successive inferior conjunctions. Whether this is accidental or dictated by orbital dynamics is a matter of debate among theoreticians.

The fraction of the surface of Venus that can be mapped at good resolution from Arecibo shows details down to between 10 and 20 kilometers in size [see illustration on page 27]. Bright areas represent enhanced radar scattering and therefore correspond to unusually rough areas. The large, extremely bright region centered at 65 degrees north latitude, five degrees east longitude is the one named Maxwell (after the physicist James Clerk Maxwell). It was first noted some years ago in early low-resolution radar observations. When Maxwell is examined in reduced-contrast images, which reveal additional detail, one can see an almost circular dark hole and a distinct "grain" running from northwest to southeast [see upper illustration on page 34]. To the west of Maxwell

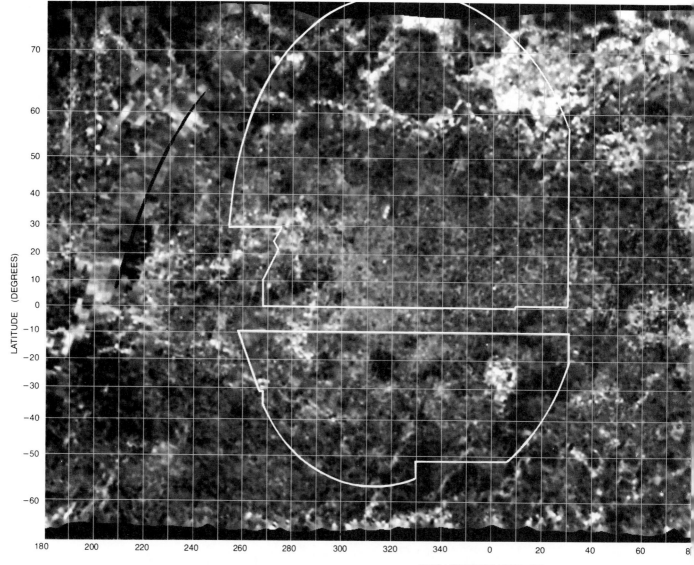

EAST LONGITUDE (DEGREES)

METER-SCALE SURFACE SLOPES ON VENUS are estimated from the same orbiter radar data used to prepare the relief map on the preceding two pages. In this presentation the darkest tones correspond to average slopes of one degree or less, the lightest areas to slopes of 10 degrees or more. The long black arc at the upper left

and the small black area to the right of it should be filled in by the end of the mission early in 1981. In general the continental regions, which are the highest in the topographic relief map, exhibit the steepest slopes. An exception is the Lakshmi plateau to the west of Maxwell. The region outlined in white is the face of Venus that has been

lies a rimmed, pear-shaped, rather dark (and therefore smooth) feature named the Lakshmi plateau (after the Hindu goddess). Because the planet itself is named for a goddess the surface features are being given women's names; the only exceptions are Maxwell and the names of two other sites that have been in use for more than a decade.

Two other bright features stand out. One, called Alpha, is on the zero meridian at about 25 degrees south latitude. The other, Beta, is centered at about 25 degrees north and 283 degrees east. A close-up of Alpha at a resolution of about five kilometers reveals a striking northeast-to-southwest grain reminiscent of the faulted basin ranges found in Nevada and adjacent areas of the southwestern U.S. To the southwest of Alpha, almost exactly on the zero meridian at 32 degrees south, lies a prominent ring-shaped feature with a bright central spot that resembles many impact craters seen on the moon and Mercury. The feature has been named Eve; the meridian that passes through the central bright spot is likely to become the planet-fixed reference origin for the official Venus longitude system.

Beta, a feature with a north-south orientation, includes a central region from which rays extend outward. Radar observations made by Goldstein show the central region to be a broad shield-shaped mountain with a small depression at the summit. R. Stephen Saunders and Michael C. Malin of the Jet Propulsion Laboratory have proposed that the feature is a shield volcano. Nearly 1,000 kilometers across, it is comparable in size to the huge volcanoes on Mars.

Just north of the equator, at about 355 degrees east longitude, a furrowed arc runs for about 1,000 kilometers. Since this feature lies almost directly under the radar beam, the intensity variation in the echo is caused mainly by large-scale tilts in the topography. The feature actually consists of two parallel ridges, each about two kilometers high, separated by a valley 90 kilometers wide. The maximum slope at right angles to the ridges approaches six degrees, an impressive tilt for such a large feature. There are no precisely analogous features on the earth, so that one can only speculate about the feature's origin.

Close examination of the radar im-

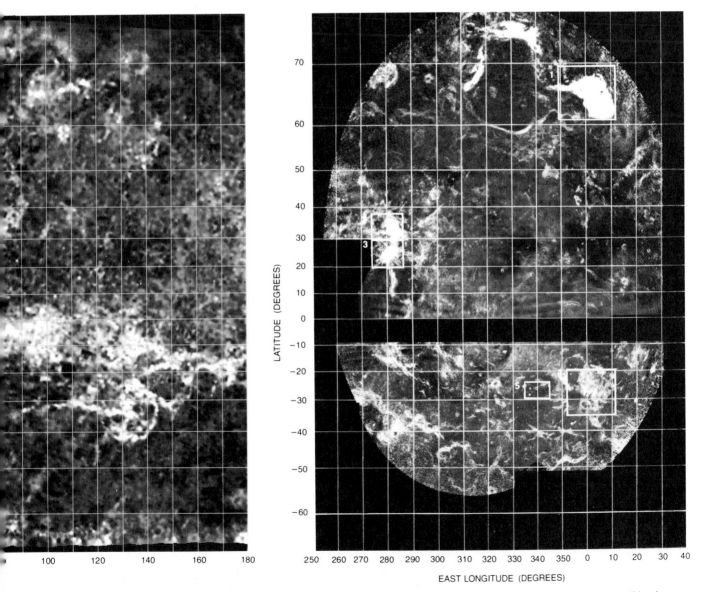

mapped by earth-based radar. At the right is a version of the earth-based-radar map at the same scale. One can see considerable similarity in the images that have been produced by the earth-based system and the orbiter one. Evidently the small-scale roughness depicted in the image made from the earth is accompanied in most places by large-scale roughness that shows up as bright in the orbiter image. Rectangles superposed on the image that was made from the earth identify five areas depicted at larger scale on the next two pages. The numbers identify Maxwell (1), Rhea Mons (2) and Theia Mons (3) in the Beta region, the Alpha region (4) and three craterlike features (5).

ages made from the earth reveals more than two dozen ring-shaped structures that are conceivably highly modified impact craters or perhaps evidence of volcanism. Barbara A. Burns of the Arecibo Observatory has plotted the number of such features as a function of their diameter and has compared the result with similar number-v.-diameter plots for other planets and the moon. For bodies with a surface visible at the wavelengths of light the distribution of craters for selected fractions of the surface agrees broadly with estimates of

the number of craters that should have been made by impacting objects over the lifetime of the solar system. Where the distribution of visible craters departs from these estimates one can infer that craters have been erased by the renewal of the surface.

Burns's analysis indicates that the surface of Venus exhibits close to the number of craters with a diameter larger than about 80 kilometers that cratering models predict should have been made within the past 600 million to one billion years. One would expect craters smaller

than about 20 kilometers to be much rarer on Venus than elsewhere because the meteorites that make such craters largely burn up in the planet's dense atmosphere before they reach the surface. This expectation cannot be tested because craterlike features smaller than 20 kilometers in diameter are below the resolution of all but an insignificant fraction of the current radar maps. The maps do show that the number of craters with a diameter between 20 and 80 kilometers are distinctly fewer than was predicted. The discrepancy may, however, simply reflect the difficulty in identifying craters of this diameter range in the radar data.

In spite of the apparent agreement with prediction for the number of large craters, the ring-shaped features observed on Venus differ in detail from typical impact craters on the moon, Mercury and Mars. One must therefore remain cautious about identifying the Venus features as impact craters. Topographic profiles of limited regions of Venus, made on the basis of radar observations from the earth, show some of the rim characteristics one would expect from an impact if one also assumes that the hot surface of the planet is capable of plastic deformation. Nevertheless, many of the rims are not quite circular, and the radial extent of small-scale roughness near them is greater than that observed for impact craters on other bodies. A better understanding of the craterlike features on Venus may come with better physical models of the mechanical properties of the hot surface and models that can predict the effects of the dense atmosphere on the patterns of the material thrown out of a crater by either impact or volcanism. Even without such models, however, the genesis of the features may be clarified by radar images made at higher resolution.

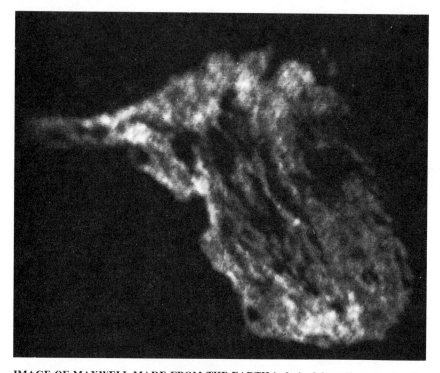

IMAGE OF MAXWELL MADE FROM THE EARTH is derived from the same Arecibo observations as those used to produce the large mosaic on page 55 but is shown here at reduced contrast to bring out interior detail. The surface resolution is about 10 kilometers. Maxwell, which measures about 750 kilometers from north to south, includes the planet's highest elevation: 11 kilometers above the planetary mean. The region exhibits a distinct grain from northwest to southeast. Dark circular "hole" at right center may be a crater partly filled with lava.

Information on the surface relief of Venus is largely missing in the radar observations of Venus made from the earth. Goldstein and his colleagues at the Jet Propulsion Laboratory have successfully applied multistation interferometry to map the relief of a few small areas of the planet, and earlier several groups determined some topography by direct radar ranging. Both types of measurement, however, have been limited to regions near the equator and have not helped to clarify the nature of the large features seen in the images of surface roughness and texture. Knowledge of a feature's relief is crucial to an interpretation of its nature and an understanding of its origin and evolution.

With this need in view the National Aeronautics and Space Administration approved the inclusion of a small radar altimeter aboard the spacecraft *Pioneer Venus 1*, placed in orbit around Venus on December 4, 1978. As the altimeter goes around the planet it measures the

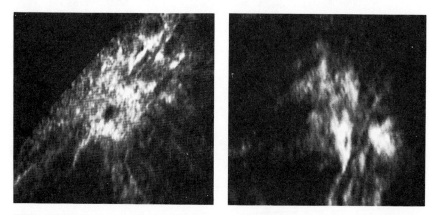

BETA REGION, also mapped by radar from the earth, contains what may be two volcanic peaks. Theia Mons (*left*) has been identified by its gently sloping circular shape as being a massive shield volcano some five kilometers high. The dark central region is presumably a volcanic caldera; the streaks radiating outward may be lava flows. (The featureless black area at the upper left has not yet been mapped.) Rhea Mons (*right*) is just to the north of Theia and may also be volcanic, but the large furrow running through its center makes interpretation uncertain.

distance to the surface below. Simultaneous tracking of the spacecraft from the earth establishes the orbit with respect to the planet's center of mass. Subtracting the spacecraft's altitude from its distance to the center of mass yields the planetary radius at the surface point observed. Although the spacecraft experiment lacks the resolution of the best observations made from the earth, it has yielded an almost global picture of the surface relief with a vertical accuracy of about 200 meters and a "footprint" resolution typically of 100 kilometers.

The *Pioneer Venus 1* mission was designed primarily for measurements of the atmosphere and ionosphere of the planet. In order to optimize these observations the spacecraft was placed in a highly eccentric orbit, ranging from a low point of about 150 kilometers above the surface at periapsis, where the atmosphere is dense enough to be analyzed directly, to a high point of more than 66,000 kilometers at apoapsis, where images of large areas of the upper atmosphere and its clouds can be obtained at light, infrared and ultraviolet wavelengths. The plane of the spacecraft's orbit is inclined at an angle of 74 degrees to the equator, so that regions of the planet poleward of latitudes 74 degrees north or 74 degrees south are inaccessible to the radar altimeter. A further limitation arises from loss of signal strength with altitude; measurements cannot be made from altitudes of more than 4,700 kilometers above the surface. Periapsis is at about 17 degrees north latitude, so that the altitude limit does not reduce coverage in the northern hemisphere. In the southern hemisphere, however, the limit is reached at a latitude of 63 degrees. Between these two parallels lies 93 percent of the surface, which becomes accessible to observation as the planet rotates slowly under the plane of the spacecraft's orbit. The spacecraft has an orbital period of 24 hours; therefore between each successive periapsis passage Venus rotates 1.5 degrees. After 243 days a complete rotation of the planet presents all longitudes to altimeter observation.

Inspection of the altimeter data from the Venus orbiter reveals a striking difference between the overall figure of Venus and the figure of the earth. As a result of the earth's substantial rate of rotation its radius at the Equator is some 21 kilometers larger than the radius at the poles. On Venus, where the rotation rate is lower by a factor of 243, there should be a negligible difference between the equatorial and the polar radii if the surface is homogeneous and has come to gravitational equilibrium. This expectation is confirmed. Eighty percent of the mapped surface falls within a height interval of two kilometers and 60 percent falls within an interval of one kilometer. Venus has several "continents " that rise two kilometers or more

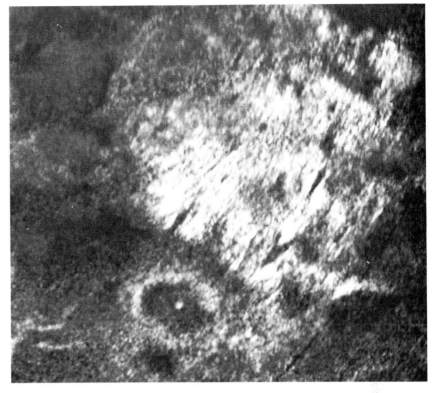

ALPHA REGION is shown with a resolution of about eight kilometers in this image made from the earth. The long parallel markings running from the northeast to the southwest resemble the faulted basin ranges that are found in Nevada and adjacent areas of the southwestern U.S. The ring-shaped feature has been named Eve. Roughly 200 kilometers across, it is probably an ancient impact crater. It has been proposed that the bright central spot serve as the zero meridian for the planet's longitude system. It lies 32 degrees south of the equator.

above the mean surface altitude (equivalent to a radius of 6,051.4 kilometers), but they constitute only about 5 percent of the total area observed. On the earth about 35 percent of the surface lies within the outer boundary of the continental shelves and 65 percent is truly oceanic.

The continental regions on the earth are not in general higher than those on Venus. A large fraction of the earth's ocean bottoms, however, lie between five and six kilometers below sea level.

When the Venus orbiting-altimeter map is compared with the roughness im-

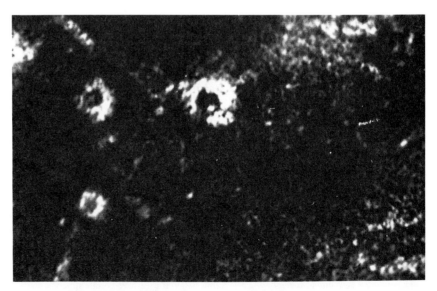

THREE RING-SHAPED FEATURES lie about 1,500 kilometers to the west of Alpha. The largest is about 100 kilometers across; the resolution is about five kilometers. Because the radial extent of rim roughness and the relief profile are unlike comparable details around impact craters on other planets and the moon, it is uncertain whether rings are really impact craters.

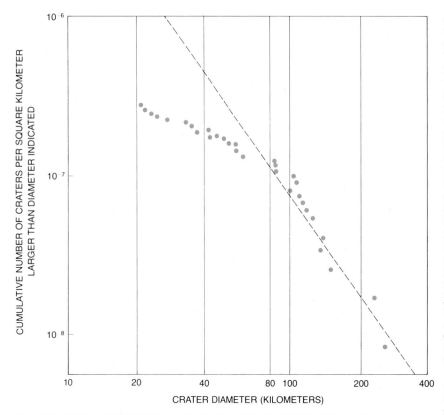

CUMULATIVE DISTRIBUTION OF RING-SHAPED FEATURES on Venus suggests that if they were created by meteorite impact, they conform to a theoretical model (*slanting line*) for the number of impact craters that would have been made within the past 600 million to a billion years. The model is derived from counts of craters on the moon, Mercury and Mars. The agreement is good for craterlike features that are larger than 80 kilometers in diameter. Objects that would give rise to craters smaller than 20 kilometers would tend to burn up in the dense atmosphere. This does not explain apparent dearth of craters smaller than 80 kilometers.

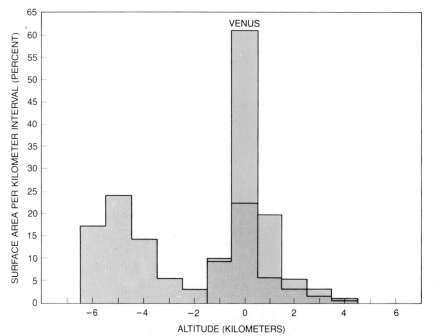

COMPARISON OF RELIEF on Venus (*color*) and the earth (*black*) demonstrates the marked contrast between the two planets. The distribution of surface height is plotted in one-kilometer intervals as a function of surface area. For Venus height is measured from a sphere with the average planetary radius of 6,051.4 kilometers. For the earth the reference is sea level. The dual nature of the earth is clear: it is 65 percent oceanic and 35 percent continental. The 65 percent outside the continental shelf is on the average about five kilometers below sea level. In contrast some 60 percent of the surface of Venus falls within half a kilometer of the average radius.

age made from the earth, a striking fact emerges: at radar wavelengths the highest regions tend also to be the roughest. Thus Maxwell, by far the brightest (that is, the roughest) large feature in the scattering image made from the earth, is seen also to be the highest, towering above the rest of the planet at altitudes of as much as 11 kilometers above the planetary mean. The Lakshmi plateau to the west of Maxwell is an exception to the rule. Although it is elevated from two and a half to three kilometers above the mean, it displays one of the smoothest large surfaces on the planet. The plateau is fringed on the north and west by fairly high mountain chains. The entire elevated continent has been named Ishtar (after the Babylonian goddess of love). The other large continental structure, called Aphrodite (after Ishtar's Greek counterpart), is along and just south of the equator, between longitudes 70 and 140 degrees east. Although Aphrodite is not as high as Ishtar, it covers nearly twice the area, a fact obscured by distortion associated with the Mercator projection.

Since Venus has continents of a sort, it is natural to wonder if, like the continents on the earth, they represent regions that are in gravitational equilibrium with their surroundings. This state, known as isostatic equilibrium, requires that topographic relief be supported by buoyancy arising from a reduced density, as an iceberg floats in water. Thus continents on the earth, which are in isostatic equilibrium, rise above the surrounding ocean basins to an extent that depends primarily on the continents' density and thickness. The degree of isostatic equilibrium on Venus can be tested by studying small perturbations in the motions of the orbiter as it passes over the large elevated structures. William L. Sjogren of the Jet Propulsion Laboratory, who has examined such perturbations in the gravitational fields over Ishtar and Aphrodite, has concluded that the regions have largely settled into a state of minimum gravitational energy.

If gravitational equilibration accounts for the large-scale relief observed on Venus, one is almost obliged to assume that substantial differences in the density of the planet's crust have been brought about by processes of differentiation. On the earth differentiation has resulted from large-scale melting of the planet's interior. Additional evidence for comparable melting on Venus is furnished by the planet's dense atmosphere of carbon dioxide with a substantial admixture of argon. Such an atmosphere argues for the thorough outgassing of a hot interior. Still further evidence for chemical differentiation on Venus is the significant amount of radioactivity measured at the planet's surface by the Russian Venera probes.

To the east of Ishtar and Aphrodite the relief is complex. The region to the east and southeast of Aphrodite is particularly chaotic; it is here that patterns suggestive of tectonic activity are most evident. A roughly circular feature 1,800 kilometers across, centered at 35 degrees south latitude and 135 degrees east longitude, may be the faint remnant of a gigantic ancient impact crater [*see illustration at right*].

In the region just north of 20 degrees south latitude and between 150 and 175 degrees east longitude is a remarkable feature that offers the strongest evidence so far for tectonic (and hence internal dynamic) activity on Venus. The feature is a long, moderately straight valley that has a sharp vertical relief approaching four kilometers. In places it shows a lateral offset, strongly reminiscent of the faults in terrestrial rift valleys.

Turning now to another aspect of the surface of Venus as it has emerged from radar observations, we have sought to correlate the time-delay spreading of the near-vertical-incidence radar echo observed by the spacecraft's altimeter with off-vertical measurements of the backscattered energy made from the earth. The former provides information about the average inclination of reflecting facets with dimensions of a meter or more. Mean values for the slope of the facets were estimated from the spacecraft's radar observations at the same time and at the same resolution as the analysis for altitude. When the results are displayed in a gray-scale pattern, they yield a map generally similar, where the maps overlap, to the centimeter-scale roughness patterns apparent in observations made from the earth [*see illustration on pages 32 and 33*]. The similarity presumably indicates that surface roughness has components ranging all the way from dimensions of a few centimeters to dimensions of many meters. A significant departure from this correlation can be seen to the northeast of Maxwell, where the centimeter-size components (the image obtained from the earth) and meter-size components (the image obtained from the spacecraft) are distributed quite differently.

From a synthesis of all the data now available there begins to emerge a picture of a planet nearly the size of the earth whose surface has been modified by all the processes that have shaped the earth's surface except erosion by rain. The tectonic motion of large crustal plates appears not to have played the dominant role in altering the surface that tectonic motion has on the earth; certainly the distribution (and probably the amount) of light continental material is far different on Venus from what it is on the earth. If the ring-shaped features prove to be the signature of ancient impacts, one must infer that much of the surface of Venus is about a bil-

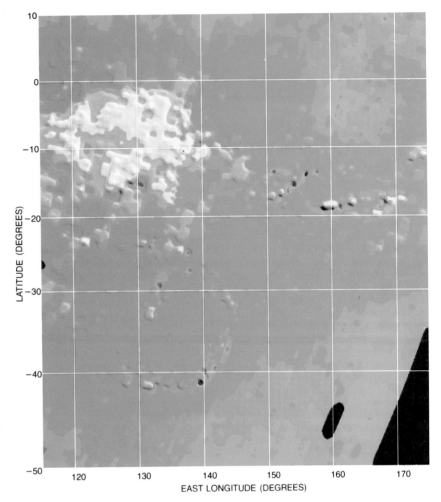

REGION EAST OF APHRODITE, based on orbiter altimetry, is depicted here with the same color contour key and shading technique used in the global map on pages 6 and 7. The region is particularly chaotic. At the right center there is a series of fairly narrow valleys flanked by elongated hills. The most likely cause of such features is tectonic motion in the crust or under it. Centered at 35 degrees south latitude and 135 degrees east longitude there appears to be a circular feature 1,800 kilometers across. It is possibly remnant of a gigantic impact crater.

lion years old. Tentatively we conclude that the planet has evolved geologically much as the earth has but not to the same degree. On the other hand, Venus has not preserved as much of its early history as the moon and Mercury have.

Questions of course remain. What are the relative roles of the three major processes—meteorite impact, volcanism and tectonics—that appear to have modified the surface of Venus? Does Venus have a central core of liquid iron-nickel as the earth does? Is there evidence on the surface of Venus for convective motion in the underlying mantle? Are volcanoes still spewing volatile substances into the atmosphere? Relatively simple radar observations have parted the veil of clouds and have given us our first comprehensive glimpse of the surface. The resolution that has been achieved so far, however, is too coarse to answer many questions in a satisfactory way. For any better understanding images with a resolution of hundreds of meters or less, such as *Mariner 9* first made of

Mars at the wavelengths of light, are needed.

NASA now has under consideration, for launching later in this decade, a spacecraft that may carry an advanced form of synthetic-aperture radar in a low circular orbit around Venus. Called the Venus Orbiting Imaging Radar (VOIR), this advanced radar system would map the entire surface of the planet at a resolution of about half a kilometer, comparable to the resolution obtained by *Mariner 9* at light wavelengths. At that resolution *Mariner 9* was able to unambiguously establish that there were both volcanism and tectonic activity on Mars, to define geologic units on the surface of the planet and to indicate their relative ages. In short, *Mariner 9* was able to provide a credible geologic history of Mars. Now that the technology for studying the surface of a planet by radar has matured, one can look forward to a similarly successful effort to understand the geologic history of Venus.

4

The Atmosphere of Mars

by Conway B. Leovy
July, 1977

It is less than a hundredth as dense as that of the earth, but it is still the principal agency altering the surface of the planet. Its winds and clouds resemble their terrestrial counterparts

The brilliantly successful series of space missions to Mars, beginning in 1964 and culminating last year in the safe touchdown of the two Viking landers, have made the planet nearly as familiar as the moon. Unlike the moon, whose story appears essentially to have ended one or two billion years ago, Mars is still evolving and changing. On Mars, as on the earth, the most pervasive agent of change is the planet's atmosphere, itself the product of the sorting of the planet's initial constituents that began soon after it condensed from the primordial cloud of dust and gas that gave rise to the solar system 4.6 billion years ago.

Although some information about the atmosphere of Mars had been gleaned from telescopic observations prior to the space flights, the information was both unverifiable and subject to misinterpretation. With the much more accurate information provided by spacecraft that have flown past Mars, gone into orbit around it and finally landed on it, we now have sufficient detailed knowledge of a second planetary atmosphere to test our understanding of atmospheric evolution and atmospheric processes previously based on the only example available: that of the earth. Except for its strikingly different composition the atmosphere of Mars behaves much like a rarefied version of our own. It transports water, generates clouds and exhibits daily and seasonal wind patterns. In response to seasonal changes in solar input localized dust storms occur and sometimes grow in strength until they envelop the entire planet. Such global-scale dust storms are a uniquely Martian phenomenon.

As a result of atmospheric weathering the primitive crystalline rock on Mars has been broken down into fine particles, oxidized and combined chemically with water to produce the characteristic reddish minerals so vividly apparent in the pictures made by the Viking landers. The mechanisms of weathering on Mars are clearly different from the dominant weathering processes on the earth, which depend on liquid water. Every-

where on Mars there is evidence of wind erosion and sedimentation. For example, the two Viking landing sites appear to have been severely scoured by winds. Deep layers of wind-blown sediment have accumulated in the Martian polar regions, and dune fields larger than any on the earth have been photographed in the region surrounding the north pole.

Because of the low atmospheric pressure on Mars and the prevailing low temperature the only stable forms of water found there are water vapor and ice. In spite of the present absence of liquid water, photographs made from vehicles in orbit show vividly that the surface of the planet has experienced fluid erosion on a grand scale. The fluid was very likely water. Running water may once have existed on the surface as a result of the melting or vaporization of subsurface masses of ice. It is also possible that at one time the climate was milder and the atmosphere was sufficiently dense and moist to have produced rainfall.

The most fundamental attributes of a planetary atmosphere are its composition and its mass, or what is equivalent, its surface pressure. As recently as 1963, the year before *Mariner 4* was launched on its successful voyage to Mars, it was believed from telescopic evidence that the atmospheric surface pressure on Mars was about 85 millibars, a value somewhat less than 10 percent of the earth's average surface pressure of 1,013 millibars. It was also believed the most abundant gas in the Martian atmosphere was nitrogen. Carbon dioxide and water had been identi-

fied spectroscopically, but both were thought to be minor components because neither one was abundant enough to contribute more than a small fraction to the assumed total of 85 millibars. Although nitrogen cannot be observed spectroscopically, it seemed the most logical principal component in view of its abundance in the earth's atmosphere.

The pressure estimate of 85 millibars was based on the intensity and polarization of reflected sunlight, using certain assumptions to separate the effects of light reflected by the planet's surface from the effects of light scattered or reflected by the atmosphere. We now know that the fraction of the reflection attributed to the gaseous component of the atmosphere was much too high because of a failure to allow for the large amount of dust and haze that was subsequently shown to be present. Although the gravity at the surface of Mars is only 38 percent of that at the surface of the earth, there is no reason in principle why Mars could not hold an atmosphere with a surface pressure nearly as high as the earth's. Hydrogen and helium are the only gases whose normal kinetic energy in the form of thermal motion is more than sufficient to enable them to escape the gravitational grip of either planet.

While *Mariner 4* was en route to its destination an analysis of better spectrographic data showed that the surface pressure on Mars had to be much lower than 85 millibars. Just how low was finally established when *Mariner 4* flew past the planet in July, 1965. From the viewpoint of terrestrial observers *Mariner 4* disappeared behind Mars and re-

EARLY-MORNING ICE FOG whitens the floors of the deep depressions in the western part of the region on Mars known as Coprates Chasma. Wispy ice clouds also drift across the surrounding mesas and valleys. The daily recurrence of such fogs suggests that water vapor condenses as ice on the ground in the night and is evaporated by the morning sun. The vapor then refreezes as it rises through the cold atmosphere. The picture is a composite of three black-and-white images made through violet, green and orange filters from the *Viking 1* orbiter in July, 1976. This picture and the one on page 36 were prepared by the Image Processing Laboratory of the Jet Propulsion Laboratory of the California Institute of Technology. These and other Viking pictures were provided by Geoffrey A. Briggs of the Viking orbiter imaging team.

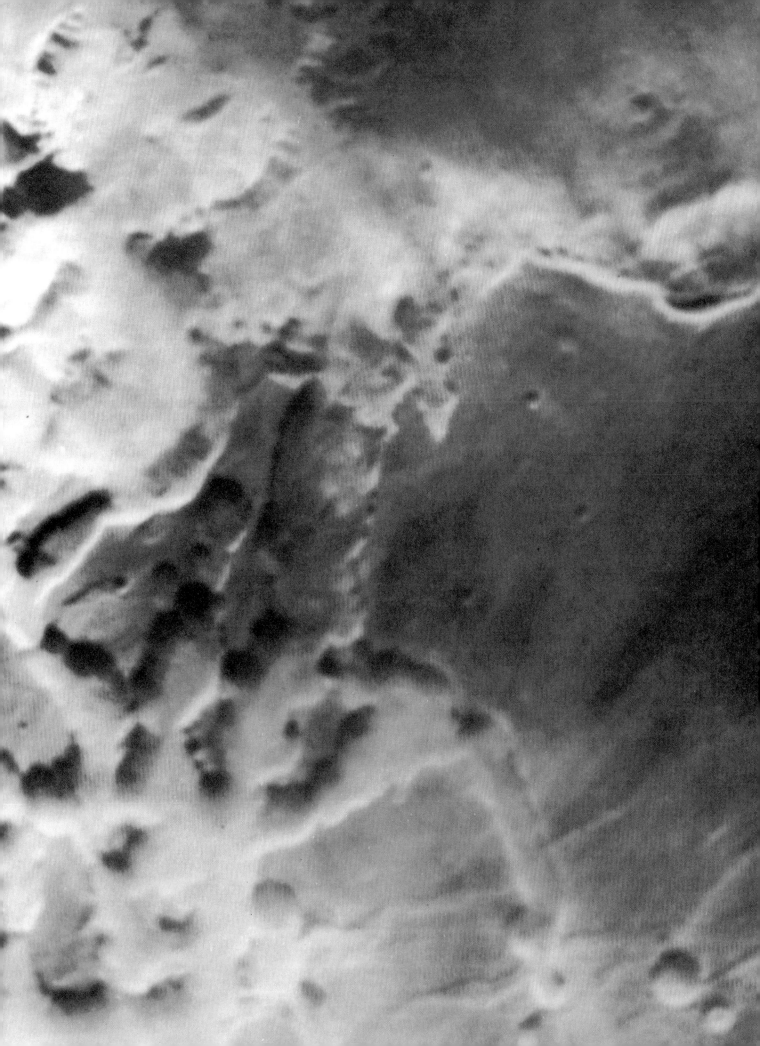

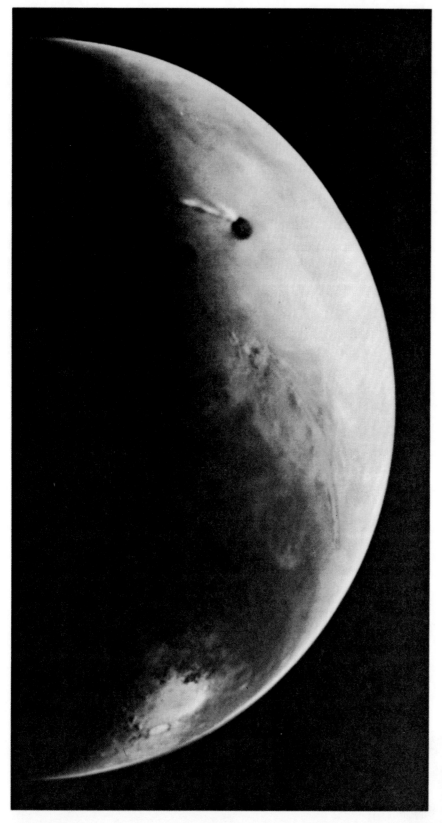

CRESCENT MARS was photographed by the *Viking 2* spacecraft in August, 1976, from a distance of 450,000 kilometers. The large circular feature to the south, the Argyre Basin, is covered with a thin frost extending from the winter polar cap. The linear features farther north are portions of the canyon system Valles Marineris, which stretches for 4,000 kilometers parallel to the equator. Farther west, near the morning terminator, the two dark circles are the enormous volcanic mountains Ascraeus Mons and Pavonis Mons. Extending westward from Ascraeus Mons are two dense cloud streamers that finally disappear into the terminator shadow 1,000 kilometers away. Cloud shadow, visible even at this range, indicates that streamers are about as high as the volcano, which is more than 20 kilometers above mean surface level, or more than twice the height of Everest. Cloud reveals a strong wind is blowing from the east.

appeared 54 minutes later. As the spacecraft's radio signals grazed the rim of the planet they were refracted by the atmosphere. The amount of refraction indicated a surface pressure of between five and seven millibars. The new spectrographic data had shown enough carbon dioxide to account for a pressure of that magnitude, so that carbon dioxide was clearly the major constituent of the Martian atmosphere. Nitrogen was relegated to the status of a minor constituent, if indeed it was present at all.

Nitrogen and the noble gases (helium, argon, neon, krypton and xenon) were measured by both Viking landers during their descent through the Martian atmosphere and subsequently on the surface of the planet. The analyses were performed by mass spectrometers, devices that first ionize a gas sample in a vacuum and then sort the resulting ions according to their mass. Two different spectrometers were used: one that operated in the near-vacuum of the Martian upper atmosphere as the lander was traversing the region between 200 and 120 kilometers above the surface and another, primarily intended for soil analysis, in which the vacuum was created artificially. Although no measurements were made between 120 kilometers and the surface, the composition of the atmosphere in the unsampled region can be inferred with considerable confidence.

The surface measurements show that carbon dioxide molecules make up about 96 percent of the total. The next most abundant gas is nitrogen, at 2.5 percent, followed by the most abundant isotope of argon, argon 40, at 1.5 percent. In agreement with earlier measurements made from the earth, oxygen molecules are present to the extent of only .1 percent. Krypton and xenon are also present in small but measurable amounts, probably along with traces of neon and helium.

On both Mars and the earth the atmosphere consists of gases released from the interior of the planet through volcanism and less violent forms of venting. Only a portion of the vented gases, however, remains in the two atmospheres as we observe them today. On the earth most of the released volatile substances are now in other reservoirs. The reservoir for most of the vented water vapor is, of course, the ocean. Most of the outgassed carbon dioxide is now tied up in calcium carbonate rocks, such as limestones, which formed on the ocean bottom. Other reservoirs of carbon, originally outgassed as carbon dioxide, are represented by deposits of coal, petroleum and oil shale and by carbon dissolved in the ocean and dispersed throughout the biosphere. The amount of carbon stored as carbon dioxide in the earth's atmosphere is insignificant.

There may also be large nonatmospheric reservoirs of vented volatile substances on Mars. Water could be tied up in caches of polar ice or widely dispersed in permafrost. Some water is known to be bound chemically in the soil. Large quantities of carbon dioxide may be adsorbed in the soil, and it is possible that some may be frozen in the small permanent south polar cap. Because the amounts of volatiles stored in such potential reservoirs are difficult to estimate, additional clues are needed to determine what the total volume of outgassed water and carbon dioxide has been. The abundance of nitrogen and noble gases in the Martian atmosphere can provide such clues.

One line of argument runs as follows. Assume that the earth and Mars were originally similar in composition, at least with regard to carbon, hydrogen and the noble gases. The noble gases do not freeze out or react chemically, and with the exception of helium they do not readily escape from the top of the atmosphere. Thus the ratio of the noble gases in the atmosphere to the total amount of carbon dioxide and water outgassed over the history of the solar system should be roughly the same for Mars and the earth. This is the "earth analogue" model.

The best noble gases to use for this comparison are krypton and the two rare isotopes of argon, argon 38 and argon 36. The more abundant isotope, argon 40, is a radioactive-decay product of potassium 40 and is therefore a better indicator of the availability of potassium than of total outgassing. Viking measurements on Mars have shown that the relative abundances of the two argon isotopes and their ratios to krypton are similar to the values measured on the earth, which is consistent with the earth-analogue model. It is risky to extrapolate the model to carbon dioxide, but if one does, one finds that the total mass of carbon dioxide that has been outgassed on Mars is only about 10 times the amount now in the atmosphere. This is a surprisingly modest quantity, and if it is even approximately correct, it indicates that the total amount of gas vented on Mars has been less than that vented on the earth by a factor of many hundreds.

A completely different way to estimate the total volume of outgassing on Mars is to examine the isotopic abundances of nitrogen, oxygen and carbon in the atmosphere to see how much the heavier isotopes of those elements have been enriched with respect to the lighter ones, which have a greater probability of escaping from the top of the atmosphere. Although neither oxygen nor nitrogen possesses enough kinetic energy in the form of thermal motion to escape from the atmosphere, both molecules can escape slowly as a consequence of being ionized by ultraviolet radiation in the upper atmosphere. When the ionized

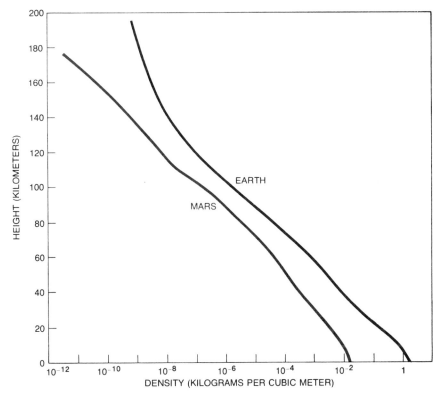

DENSITY OF THE MARTIAN ATMOSPHERE is only about a hundredth the density of the earth's atmosphere at ground level. Up to a height of about 100 kilometers the density of the earth's atmosphere decreases more rapidly than the density of Mars's. Above 100 kilometers the situation reverses. The density profile of the Martian atmosphere below 100 kilometers is derived from an analysis of the performance of the *Viking 2* lander as it traversed the atmosphere on its way to the surface. Analysis was done by Alvin Seiff and Donn B. Kirk of Ames Research Center of National Aeronautics and Space Administration. Above 100 kilometers atmospheric density is inferred from an analysis of data supplied by entry mass spectrometer of *Viking 2* lander. The analysis was done by Alfred O. C. Nier of University of Minnesota.

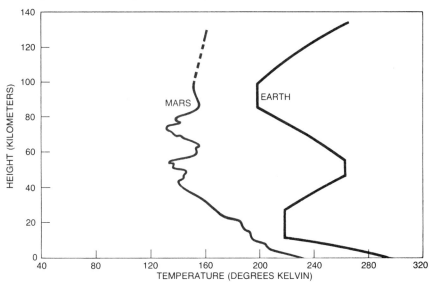

TEMPERATURE OF THE MARTIAN ATMOSPHERE departs markedly from the pattern in the earth's atmosphere. The one common characteristic of the two atmospheres is a steady drop in temperature up to a height of about 10 kilometers. Below that height both atmospheres are warmed by absorbing energy from the ground, which is directly heated by the sun. The temperature peak in the earth's atmosphere near 50 kilometers is caused by the absorption of ultraviolet radiation by ozone. There is no analogous ozone layer in the Martian atmosphere to absorb energy. The earth's atmosphere above 100 kilometers is also heated by ultraviolet radiation. Comparable heating is again absent in the atmosphere of Mars. Irregularities in the temperature profile of the Martian atmosphere between 50 and 100 kilometers are probably due to the expansion cooling of layers of rising gas and the compression heating of alternate layers of sinking gas. Such layers may be produced by a global system of tidal winds. Martian temperature near 130 kilometers was obtained from a Viking entry experiment analyzed by William B. Hanson of University of Texas. Remaining temperatures are from analyses by Seiff and Kirk.

molecules and electrons recombine, the energy released is more than enough to dissociate the molecules into their constituent atoms. The energy left over imparts enough kinetic energy to the atoms of oxygen and nitrogen so that both gases can slowly escape. The same process does not operate on the earth because of the earth's stronger gravity.

The Viking measurements show that the various isotope abundances remain constant up to 120 kilometers. Above that altitude the ratios begin to change because turbulence is insufficient to keep the gases well mixed. In the ultrararefied atmosphere above 120 kilometers the concentrations of the various gases are controlled by molecular diffusion, a process that enables each gas to distribute itself according to its molecular mass, the lighter gases rising to the top like cream and ultimately escaping into space. The zone that separates the well-mixed lower region of the atmosphere from the diffusion region is called the turbopause. On Mars the turbopause is at about 120 kilometers, where the density of the atmosphere is only about a hundredth the density at the terrestrial turbopause.

Outward escape from a planetary atmosphere is possible only for molecules above the turbopause, because it is only there they can escape before colliding with other molecules. Since lighter isotopes are relatively more abundant above the turbopause, they will escape more rapidly than their heavier counterparts. If continued long enough, this process will lead to selective enrichment of the heavy isotopes with respect to the light ones throughout the atmosphere. The Viking measurements show that the ratio of the rare heavy isotope of nitrogen, nitrogen 15, to the common isotope, nitrogen 14, is larger than the corresponding terrestrial ratio by about 60 percent. The degree of enrichment depends on the total amount of nitrogen vented, when it was vented and the height of the turbopause. The higher the turbopause, the less the enrichment of nitrogen 14 with respect to nitrogen 15 in the region from which escape can occur. Enrichment also depends on such external factors as the flux of solar ultraviolet energy, which facilitates the escape process by imparting the necessary energy of ionization. Models of the evolution of the Martian atmosphere based on the ratio of nitrogen isotopes suggest that the outgassing of nitrogen began early in the planet's history and that the total volume vented has been large, perhaps 100 times the amount now present in the atmosphere.

According to the Viking measurements, there has been no comparable enrichment in the heavier isotopes of carbon and oxygen. How can that be explained? One proposal is that the atmospheric carbon and oxygen, unlike the atmospheric nitrogen, are in contact with large surface or subsurface reservoirs of carbon- and oxygen-containing materials, so that there can be atomic exchanges between the atmosphere and the reservoir. The amounts of carbon and oxygen in the atmosphere may represent only the tip of the iceberg. Thus the absence of observable enrichment in the heavier isotopes of carbon and oxygen can be used as an argument for a large volume of outgassing having created a large reservoir of carbon dioxide. It seems safe to conclude that the outgassing of Mars, although it was far less efficient than the outgassing of the earth, produced substantially more carbon dioxide and water vapor than is now found in the atmosphere.

Unlike nitrogen and the noble gases, water vapor is a highly variable component of the Martian atmosphere. Earth-based measurements had shown a seasonal cycle and a suggestion of a daily one. Spectrometers aboard the Viking orbiters have now shown that the amount of water vapor varies strongly with latitude. Although the vertical distribution of water vapor is difficult to establish spectroscopically, the total mass of water per unit of area of a column extending from the ground to the top of the atmosphere is readily calculated. By converting the amount of water vapor in such a column to its liquid equivalent one obtains the number of "precipitable centimeters": the depth of the liquid if it were precipitated.

The values measured so far by the Viking orbiters range from less than 10^{-4} precipitable centimeter in high southern latitudes in midwinter to 10^{-2} precipitable centimeter in high northern latitudes in midsummer. The earth's atmosphere normally contains two or three precipitable centimeters, so that one's first impression is that the atmosphere of Mars is exceedingly dry. A fairer comparison, however, is one made with the total amount of water in the earth's atmosphere above the level at which its pressure matches the surface pressure on Mars: seven millibars. By this criterion the atmosphere of Mars is very wet; even the lowest value measured on Mars is several times larger than the water content of the earth's atmosphere. A more relevant measure is the relative humidity, or degree of saturation of the atmosphere. On the average the relative humidity of the Martian atmosphere is so high that the abundance of water vapor seems to be controlled largely by saturation. The atmosphere is about as wet as it can be for the prevailing atmospheric temperature.

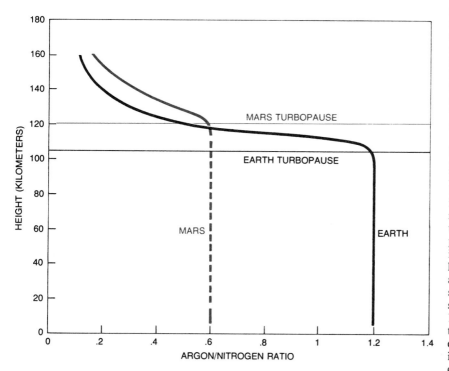

RATIO OF ARGON TO NITROGEN in the atmospheres of Mars and the earth declines in parallel above about 120 kilometers but exhibits sharply different values below the respective turbopauses of the two planets. The turbopause marks the transition zone between the lower region of the atmosphere, where winds keep the component gases well mixed, and the low-density upper region, where components can become separated by molecular diffusion. Above the turbopause the argon/nitrogen ratio falls sharply as nitrogen, the lighter of the two gases, increases in relative abundance prior to its slow escape from Mars. Martian profile above 110 kilometers is derived from Viking entry experiments. Values near Martian surface are from mass spectrometer on *Viking 2* lander in an experiment by Klaus Biemann of Massachusetts Institute of Technology and Tobias C. Owen of State University of New York at Stony Brook.

Factors other than saturation, however, are clearly needed to maintain the observed variation in water-vapor content with latitude. One important factor is the distribution of water-vapor sources and sinks. In late summer at each pole there is a small residual ice cap, and at the north pole, at least, the cap is known to be water ice. Sublimation of the ice under the summer sun would provide an abundant source of water vapor. Not surprisingly, the largest amounts of vapor observed by Viking are around the edges of the north polar cap. Curiously, however, no similar enhancement was detected around the residual midsummer south polar cap by the *Mariner 9* orbiter. If the Viking orbiters succeed in gathering data for a full annual cycle, the mystery may be cleared up

Water vapor is the key ingredient in the chemistry of the Martian atmosphere. Unlike the earth, which is shielded from solar ultraviolet radiation by a dense layer of ozone, Mars is exposed to this energetic radiation right down to the surface. As a result throughout the atmosphere carbon dioxide is dissociated into carbon monoxide (CO) and atomic oxygen (O), and water vapor is dissociated into atomic hydrogen (H) and the hydroxyl radical (OH). The H and OH are extremely reactive. For example, they can catalyze the recombination of CO and O back to carbon dioxide, which helps to explain why the atmosphere consists mainly of carbon dioxide. Atomic hydrogen and molecular oxygen combine through a series of reactions to produce hydrogen peroxide (H_2O_2), a powerful oxidizer, which may have an important influence on the chemistry of the soil.

A small amount of molecular hydrogen (H_2) is produced by the dissociation of water vapor. Being quite stable, H_2 is well mixed below the turbopause. Above the turbopause it dissociates into atomic hydrogen, which slowly escapes. Extrapolated over the lifetime of Mars (4.6 billion years) the current estimated rate of escape could account for the loss of a layer of liquid water less than a meter deep. That is probably much less than the total amount of water vented from the planet.

Although the four hydrogen-containing species, H, H_2, OH and H_2O_2, are chemically important, they are below the threshold of detection by the Viking instruments. The atmospheric chemical events I have described are inferred from terrestrial experience. Ultraviolet spectrometers on the Mariner spacecraft, however, have detected the atomic hydrogen that is escaping from the upper atmosphere of Mars. These observations lend support to the presumed dissociation of molecular hydrogen above the turbopause.

Stronger support for the inferred

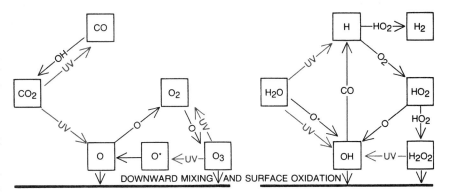

ATMOSPHERIC CHEMICAL REACTIONS are initiated when carbon dioxide (CO_2) and water vapor (H_2O) are dissociated by ultraviolet radiation (UV). In the oxygen cycle (*left*) CO_2 breaks up into carbon monoxide (CO) and atomic oxygen (O). The latter recombines to form O_2 and ozone (O_3). Both O and O_3 contribute to the oxidation of rocks and soil. In the hydrogen cycle (*right*) H_2O dissociates into atomic hydrogen (H) and the hydroxyl radical (OH). Reaction of H with O_2 produces the perhydroxyl radical (HO_2), which can react with itself to form hydrogen peroxide (H_2O_2). Both H_2O_2 and OH can contribute to surface oxidation. The two cycles interact mainly through the reactions of CO with OH, O_2 with H, and O with HO_2. Excited atomic oxygen (O'), formed by ultraviolet dissociation of O_3, can also break H_2O up into OH radicals. Molecular hydrogen (H_2) is also formed in the hydrogen cycle. Both H_2 and O_2 are slowly removed from the atmosphere by upward mixing and eventual escape. Quantitative aspects of the cycles depicted here have been worked out by Michael B. McElroy of Harvard University and Thomas M. Donahue of University of Michigan and independently by Truman D. Parkinson and Donald M. Hunten of the Kitt Peak National Observatory.

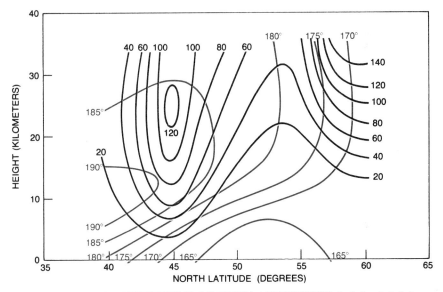

TEMPERATURE OF WINTERTIME MARTIAN ATMOSPHERE (*color*), plotted in a meridional plane, has been derived from infrared measurements made by *Mariner 9*. The temperatures, given in degrees Kelvin, can be used to infer the distribution of the east-to-west component of the winds (*black*), given in meters per second. (One hundred meters per second equals 224 miles per hour.) Wind velocities shown are in addition to surface winds, which are typically weak in winter. *Mariner 9* infrared experiment from which temperatures were derived was by Rudolf A. Hanel, B. J. Conrath and co-workers at Goddard Space Flight Center of NASA.

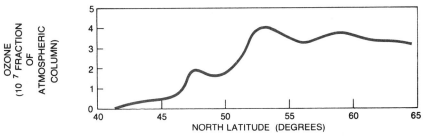

WINTERTIME OZONE DISTRIBUTION is plotted for the same plane as that used for temperatures and wind velocities in middle of page. Vertical axis shows the number of ozone molecules as a fraction of all molecules in a vertical column of atmosphere. Ozone increases toward pole, where temperatures and water-vapor concentrations are low. Values were obtained by Charles A. Barth and collaborators at University of Colorado from *Mariner 9* ultraviolet data.

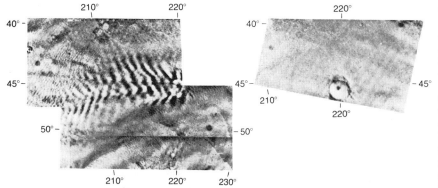

WAVE CLOUDS IN THE ATMOSPHERE OF MARS in midwinter are clearly visible in the mosaic of three pictures at the left, made by *Mariner 9* in 1972. The long train of clouds almost obscures the large crater over which the clouds have formed. The individual wave clouds are about 20 kilometers apart, or about half the diameter of the crater. The cloud-free crater can be seen at the lower edge of the single picture at the right, which was made just a day earlier. Latitudes and longitudes are given along the borders of the mosaics. In order to keep the apparent relief of the crater and the other surface features from reversing, the pictures are reproduced with north at the bottom, so that the sun illuminates the surface features from the top.

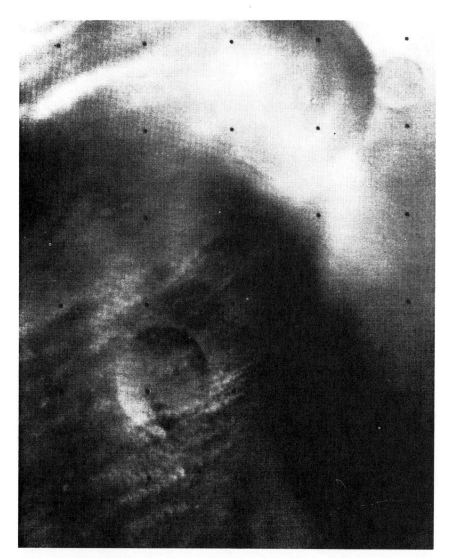

WAVE CLOUDS IN THE LEE OF A MARTIAN CRATER were photographed by *Mariner 9* in 1972. The broad, diffuse wave cloud that appears at the upper right is similar to the clouds visible above the 40-kilometer crater in the low-resolution mosaics at the top of the page. The clouds are probably water ice. The crater that gives rise to this cloud is only about seven kilometers in diameter. To the lower left and downwind from crater are cloud streamers whose sharp edges and small-scale structure suggest they are composed not of water but of solid carbon dioxide, which can freeze out of the atmosphere rapidly over polar regions in winter.

chemistry is provided by the observed distribution of ozone (O_3) in the atmosphere of Mars. Although the ozone is much less abundant than it is in the atmosphere of the earth, there are readily detectable concentrations of it over the Martian polar regions in winter. At lower and warmer latitudes it is usually undetectable. Ozone is produced by the combination of O and O_2, which have been generated by the decomposition of carbon dioxide. Ozone and its precursor, atomic oxygen, are destroyed by catalytic reactions that depend on H and OH. Thus the ozone concentration will be low wherever H and OH are in good supply, which they are wherever the source material, water vapor, is abundant. Since water vapor is virtually absent around the winter pole, ozone is plentiful there. The limitation the dissociation of water vapor places on ozone concentration plays a similar role in the earth's atmosphere. Above an altitude of about 40 kilometers the ozone concentration is limited by the same catalytic reactions with H and OH. Hence the Martian atmosphere resembles the earth's stratosphere in chemistry as well as in pressure.

Since the Martian atmosphere is close to saturation, one would expect ice clouds to be plentiful, as indeed they are, at least during the northern summer season the Viking vehicles have examined so far. As one might also expect, the clouds of water ice are thin, with diffuse edges. There is nothing resembling the dense, sharp-edged cumulus clouds we know on the earth. The Martian clouds are of four general types: convective clouds, wave clouds, orographic clouds and fogs.

Convective clouds are formed as gas near the surface is heated during the day, rises and cools by expansion. When the temperature falls to the saturation point, clouds form in distinct puffs of nearly uniform size and spacing. Such cloud patterns are frequent at about noon over equatorial upland regions. The spacing of the puffs and the cloud heights estimated from the locations of their shadows suggest that active convection is taking place in a layer from five to eight kilometers deep, which is somewhat greater than the depth of the diurnally heated layer over continental areas on the earth. The powerful daily stirring by convection ensures that all the atmospheric constituents of the layer can interact efficiently with the Martian surface.

The best pictures of wave clouds were made by *Mariner 9*, which examined the northern hemisphere in winter. Wave clouds form when a strong wind blows steadily across an obstacle such as a high ridge. Trains of atmospheric waves resembling water waves form in the lee of the obstacle. If moisture and temperature conditions are right, the crests of the waves are marked by clouds that appear to be stationary.

The wave clouds on Mars show that in winter the meteorology of the middle-latitude regions bears a striking resemblance to that of similar regions on the earth. The prevailing winds are westerly at all levels up to a considerable altitude, and the strongest winds, like terrestrial jet streams, are found several kilometers above the surface. By combining remotely sensed temperature information with wave-cloud photographs one can reconstruct a three-dimensional picture of the wind distribution on particular days. Typically the wind blows from the west with speeds of 10 to 20 meters per second near the surface to more than 100 meters per second (224 miles per hour) near the level of the jet stream, above 10 kilometers. These speeds are about twice as great as those that are found at the corresponding levels on the earth.

Orographic clouds form when atmospheric gas is forced to rise slowly and steadily as it moves up the slope of a large upland. Such clouds are common over the higher volcanic regions of Mars. In the northern summer season, when the atmosphere is moist, large areas are covered with rather uniform thin clouds, which can be observed from the earth. For days on end their pattern of growth and decay is repeated with remarkable precision, indicating the regularity of the day-to-day weather.

One of the most surprising atmospheric observations made by the Viking orbiters was early-morning ground fogs in several low-lying areas. The fogs are evidently created when ground frost is converted into vapor by the morning sun. The vapor condenses again as it rises through the chill atmosphere of the early morning. Since the fogs are seen day after day in the same locations, it seems likely that the atmospheric water vapor forms a new frost on the surface every night.

Water is not the only substance capable of forming clouds on Mars. In the winter polar regions and at high altitudes the temperature can fall low enough for carbon dioxide to condense, creating clouds of dry ice. Since the condensation and sublimation of carbon dioxide can be rapid, dry-ice clouds, unlike water-ice clouds, can have sharp edges. Several bright, sharp-edged clouds photographed by Mariner and Viking spacecraft are suspected of being dry-ice clouds. It is also possible that dry-ice snowstorms may occur in winter and help to create the seasonal dry-ice polar caps. Much of the dry ice in the caps, however, is probably deposited when carbon dioxide gas comes in direct contact with the frigid Martian soil and condenses.

About 20 percent of the carbon dioxide in the Martian atmosphere is cycled between the atmosphere and the winter

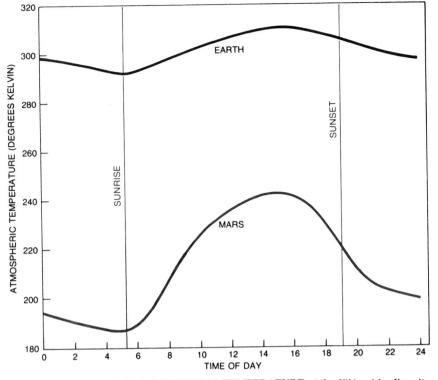

DAILY VARIATIONS IN ATMOSPHERIC TEMPERATURE at the *Viking 1* landing site (*color*) are qualitatively similar to those at China Lake, Calif., a desert site (*black*). In both cases the temperature touches a minimum around sunrise and reaches a peak about 10 hours later. The daily range, however, is about three times greater on Mars than it is on the earth. At Viking site range is 55 degrees, from about 187 to 242 degrees Kelvin (−86 to −31 degrees Celsius). At China Lake range is 18 degrees, from 292 to 310 degrees K. (19 to 37 degrees C.).

polar cap each season, causing a corresponding variation in the atmospheric pressure. The pressure variation, which is felt throughout the atmosphere, not just in the polar-cap regions, has been measured by the barometers on both Viking landers. The magnitude of the pressure change agrees remarkably well with pre-Viking theoretical models of the carbon dioxide exchange between the polar caps and the atmosphere. The measurements show that any other large reservoirs of carbon dioxide, for example carbon dioxide adsorbed on particles of Martian soil, do not exchange seasonally with the atmosphere. There may, however, be a large reservoir of adsorbed carbon dioxide capable of exchanging with the atmosphere on a much greater time scale.

Winds on Mars have long been a subject of speculation. In recent years serious efforts have been made to develop theoretical models of them with techniques similar to those applied to the forecasting of terrestrial weather and climate. The techniques should be applicable since both planets have a similar sequence of seasons (because their polar axes are tipped almost identically) and a nearly identical daily rotation rate (24 hours 37 minutes for Mars). The rotation rate is important because it determines the magnitude of the deflection of moving masses of at-

mosphere. The deflection, or Coriolis force, is responsible for the counterclockwise rotation of winds around low-pressure areas in the northern hemisphere and the clockwise rotation around low-pressure areas in the southern hemisphere.

The picture that emerges from the theoretical investigations is a planetary wind pattern with two major regimes. One is a middle-latitude winter regime similar to such a regime on the earth, with prevailing westerlies, high-altitude jet streams and traveling disturbances, or storms. The other is an equatorial summer regime free of traveling disturbances, in fact free of all day-to-day variations except the gradual ones associated with the slow march of the seasons. Under the second regime the main influence on the wind flow is the daily variation in solar heating and its interaction with local topography. Thus the heated gas should move up regional slopes during the day and down again during the night. Since there are terrain slopes almost everywhere on Mars, slope winds, deflected by planetary rotation, should be widespread.

In addition the daily heating cycle should create atmospheric tides, or planetwide oscillations in the wind pattern. Since Mars is a desert planet, the tides should be stronger than those on the earth. Near the surface the tidal winds are weak: on the earth they attain at most one meter a second; during the northern hemisphere summer season on Mars they should reach perhaps five or 10 times that velocity. The velocities increase sharply, however, with altitude. On the earth they reach about 50 meters per second between altitudes of 80 and 100 kilometers. Although the corresponding velocities on Mars have not been measured, one can infer from the height of the turbopause that they are higher. The tidal winds are effective stirring agents in the upper atmosphere, and so they probably control the height of the turbopause on both planets. Since the Martian turbopause is at a lower density level than the turbopause on the earth, as a consequence of more efficient stirring by winds, it seems likely that the atmospheric tides on Mars are stronger than those on the earth.

Do Viking observations provide any other support for this conjecture? Perhaps the best evidence for the strength of the atmospheric tides on Mars is supplied by barometer readings at the two Viking landing sites. The instruments show large daily variations in atmospheric pressure, a clear indication of strong atmospheric tides. Supporting evidence was provided by measurements of vertical temperature profiles made by the entry vehicles. At higher levels, where the tidal winds should be strong, there should be large tempera-

ture excursions caused by alternating layers of atmospheric gas heated by compression and gas cooled by expansion. Theory predicts that the excursions will be about 20 kilometers apart and that their amplitude will increase with height.

The temperature profiles recorded during the entry of the Viking lander vehicles show temperature excursions with just these properties. Hence it does appear that on Mars, as on the earth, the height of the turbopause is set by the strength of the atmospheric tides. The turbopause in turn acts as a kind of valve regulating the escape of light gases. Because the Martian turbopause is relatively high the volume of light gases available for escape is correspondingly low, and it is partly for this reason that the current escape rate of hydrogen on Mars is so low.

The great dust storms of Mars were observed long before the age of space exploration. Because of the low density of the Martian atmosphere much stronger winds are needed to raise dust storms on Mars than are needed on the earth. A wind velocity as low as six or seven meters per second (13 to 16 miles per hour) can begin to raise sand grains in many terrestrial deserts. Laboratory studies show that winds of at least 30 to 60 meters per second (65 to 135

miles per hour) are required on Mars. Since there is no rain on Mars to clear the air, small dust particles can remain suspended for weeks or months.

Winds strong enough to cause dust storms develop in winter in the snow-free belt adjacent to the edge of the Martian polar cap. Several dust storms were photographed in that region by both *Mariner 9* and the Viking spacecraft. Far more impressive are the global dust storms, one of which totally blanketed the planet when *Mariner 9* arrived and went into orbit in November, 1971. Prior to the Viking missions the global storms were believed to occur regularly, once each Martian year, at about the time the planet makes its closest approach to the sun. Very recently, however, the Viking spacecraft have detected dust storms on a nearly global scale at a much earlier Martian season. Evidently the very large storms are commoner than has been thought. Beginning with a small-scale storm (or several such storms) in the southern tropics, the global storms spread rapidly, so that within a few days they cover most of the planet.

What can generate the strong winds of such huge storms? It must be significant that Mars's orbit is considerably more elliptical than the earth's. The amount of solar energy impinging on Mars is about 40 percent greater when the planet is closest to the sun than it is when the planet is farthest away. The corresponding figure for the earth is only 3 percent. This fact alone, however, cannot account for the wind velocities required.

A number of explanations have been advanced, but I shall discuss only the one I find most plausible. The strength of the global tidal winds is very sensitive to the heating of the atmosphere. That heating in turn depends not only on the amount of solar energy reaching the Martian surface but also on the dustiness of the Martian atmosphere: the dust particles absorb sunlight and can heat the surrounding gas directly and efficiently. If the atmosphere is sufficiently laden with dust, the global tidal winds can grow in intensity until they are nearly strong enough to raise dust by themselves. Presumably this self-regenerative process begins in localized areas, where the tidal winds and local topographic winds combine to produce velocities sufficiently high to put a critical amount of dust into the local atmosphere. As the dust is spread by the winds it augments the heating of the atmosphere over a large area, giving rise to tidal winds strong enough to raise dust over still wider areas.

It seems clear that suspended dust influences the behavior of the atmosphere on Mars beyond anything known on the earth. Even during the season of the recent Viking landings, a season in which the atmospheric dust load was fairly light, the sky had a pinkish-yellow cast

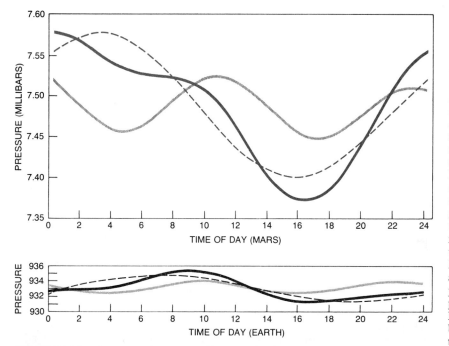

DAILY VARIATIONS IN ATMOSPHERIC PRESSURE, in relation to total pressure, are about five times greater at the *Viking 1* landing site than they are at China Lake on the earth. The solid curve in color at the top shows that in the course of a Martian day the atmospheric pressure varies by .2 millibar, which is about 2.5 percent of the total mean pressure of 7.5 millibars. The curve can be decomposed into a daily harmonic (*broken curve*) and a semidaily harmonic (*light-color curve*). The black curve below it is the daily pressure variation at China Lake, which is similarly decomposed into two harmonics. To match the daily variation on Mars a barometer on the earth would have to vary by some 23 millibars in the course of a day. Actual range at China Lake is only 4.5 millibars. Large daily fluctuations on Mars indicate that atmospheric tides and accompanying wind systems are much more intense there than on the earth.

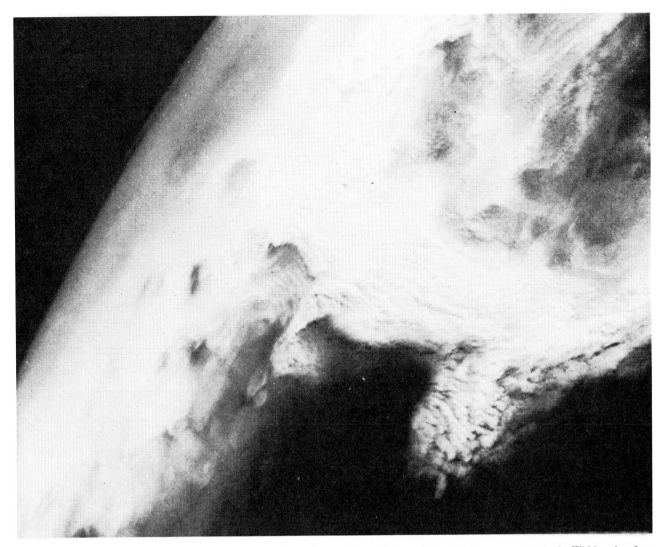

VAST MARTIAN DUST STORM was photographed by the *Viking 2* orbiter in February of this year from a point 33,000 kilometers above the surface of the planet and at 51 degrees south latitude with a camera pointed to the west-northwest. The center of the image is at 42.5 degrees south latitude and 108 degrees west longitude, which places the storm in the region near Claritas Fossae and Thaumasia Fossae between 20 and 45 degrees south latitude. Within a few days the storm had spread over most of the southern hemisphere and then lasted for several weeks. When the storm began, it was early spring in the Martian southern hemisphere. Before these observations were made by the Viking spacecraft it was believed such near-global dust storms were limited to the season near the southern summer solstice.

in pictures taken from the landers—evidence of a substantial amount of light-scattering dust in the atmosphere. It also seems necessary to invoke the solar heating of a dust-laden atmosphere to account for daily variations in surface pressure as large as those measured at the landing sites. Thus there seems to be an efficient positive-feedback system whose components are winds, dust and atmospheric heating.

Although uncertainties remain, the present picture of processes operating in the atmosphere of Mars is remarkably detailed in spite of the brief history of the closeup exploration of the planet. The Mariner and Viking pictures and measurements have disclosed many striking similarities between the Martian atmosphere and the one we know on the earth. Chemistry, dynamical processes and atmospheric evolution are linked closely on both planets. The effort to understand these interacting processes on another planet is helping to add perspective and breadth to studies of our own planet.

Can we yet say anything about the Martian climate in the past? In particular, do the great channels and other apparent manifestations of erosion by water indicate that Mars once had an extended period of warm and wet weather? I believe the weight of the evidence makes it unlikely. It appears that the total amount of outgassing on Mars has been far less than that on the earth. Even if the total vented mass were 100 times the present atmospheric mass or more, the availability of water vapor in the atmosphere may not have been much greater then than it is now. The limitation on the amount of water vapor in the Martian atmosphere is the low temperature. Even an atmosphere approaching the density of the earth's composed chiefly of carbon dioxide and nitrogen could not have been much warmer than the present one, and so it could not have held much more water vapor.

Conceivably the sun itself was hotter when water eroded the features we now see on Mars, but it seems more likely that these features resulted from cataclysmic events such as meteorite impacts or volcanic explosions. The Martian crust has probably contained large amounts of water, either frozen or chemically bound, for a long time, so that any sudden heating could have released local floods. Such a cataclysm could also have produced brief, intense local storms in which the sudden vaporization of water was quickly followed by torrential rains. There may have been similar cataclysmic floods in very early stages of the earth's history, before there were oceans. It appears that more definite answers to the questions of past Martian climates and the origin of the channels will have to await the next stage in the exploration of Mars.

5

The Surface of Mars

by Raymond E. Arvidson, Alan B. Binder
and Kenneth L. Jones
March, 1978

*The Viking spacecraft have provided an unparalleled
view of it from orbit and from the ground, adding much
evidence on how it has been shaped by volcano,
meteorite impact, water and wind*

The two Viking landers have now been on the surface of Mars for nearly a full Martian year; the two Viking orbiters have been circling and photographing the planet for slightly longer. Ever since the landers touched down in the summer of 1976 they have been gathering data on the characteristics of the Martian atmosphere, rocks and soil. Meanwhile the orbiters have been monitoring the water-vapor content of the atmosphere, mapping the temperature of the surface and photographing the surface with unprecedented resolution and clarity.

Together the photographs made and analytical experiments done by the four spacecraft reveal that Mars is a planet with an even more complex history than had been suspected. They have provided evidence that even the most ancient cratered terrain has been modified by volcanic activity, that early in the planet's history flowing water was a significant agent in shaping its features and that since then surface material has been extensively redistributed by high-velocity winds. Surprisingly, the surface has been little eroded by such aeolian activity. In appearance the surface of Mars is more like rocky volcanic deserts on the earth than it is like the highly cratered surface of the moon, yet Mars, once visualized as being largely a world of gently rolling dunes, seems to possess little sand. The Viking mission has also provided evidence that has both strengthened and altered hypotheses concerning the early Martian atmosphere and climate proposed after the *Mariner 9* mission of 1971 and 1972.

Mars before Viking

Mariner 9 photographed practically the entire surface of Mars at a resolution of a few kilometers and a small fraction of the surface at a resolution of a few hundred meters. The global coverage of Mars by *Mariner 9* indicated that the planet was divided into two distinctly different hemispheres: a rugged, heavily cratered southern hemisphere traversed by huge channel-like depres-

sions and a smooth, lightly cratered northern hemisphere dotted with extinct volcanoes. The two types of terrain are divided roughly along a great circle inclined to the equator by about 30 degrees.

The craters in the southern hemisphere range in size up to the basin Hellas, which is 1,600 kilometers across. The abundance of craters in some of the more heavily cratered regions is comparable to the abundance of craters in the bright highlands of the moon. Samples of rocks and soil collected on the lunar highlands during the Apollo missions have been dated, and their ages indicate that most of the craters on the lunar highlands were excavated between four and four and a half billion years ago. At that time the moon was torrentially bombarded by the interplanetary debris left over from the formation of the solar system. Since the abundance of craters on Mars is similar to that on the moon, it is believed the Martian cratered terrain is approximately the same age as the lunar highlands. In other words, almost half of the surface of Mars is ancient terrain where many of the larger land forms have remained unchanged for some four billion years.

Mariner 9 revealed that the sparsely cratered areas in the northern hemisphere of Mars are plains of lava that extensively flooded the surface at different times after the heavy bombardment of the planet ceased. The few craters on the plains record the occasional impact of a stray asteroid or comet. Although there is no way as yet to estimate with confidence the absolute ages of the plains, the abundances of craters in different regions vary widely, which implies that the ages of the plains range from hundreds of millions to some billions of years.

The channels on Mars photographed in such detail by *Mariner 9* seemed to imply that the past climate of the planet must have been quite different from the present one. If all the water in the Martian atmosphere today were condensed in one place, it would form a body of water no larger than Walden Pond. In

fact, the abundance of water in the atmosphere today is so low that rain is an impossibility.

By the time the Viking landers touched down most investigators were convinced that the largest channels, which are tens of kilometers wide, were carved by water derived from the melting of ice below the surface. Some investigators believe the ice is a remnant from a denser atmosphere the planet may have had in its first billion years. According to calculations made by Fraser P. Fanale of the Jet Propulsion Laboratory of the California Institute of Technology, the early Martian atmosphere may have contained ammonia and methane in addition to carbon dioxide and water vapor. According to James B. Pollack of the Ames Research Center of the National Aeronautics and Space Administration, such an atmosphere would have acted to trap infrared radiation (the "greenhouse effect") and would have been warm enough to hold substantial amounts of water vapor. At some time in the past various reactions would have removed the ammonia and methane from the atmosphere. The atmosphere would then have become more transparent to infrared radiation. As a consequence the atmosphere would have cooled and the water vapor in it would have condensed and precipitated onto the surface, where it would have made its way to the polar regions and into the planet's impact-fractured crust and regolith (the loose material lying on the crust).

Mariner 9 also revealed that both the north and south poles of Mars are covered with ice and windblown dust as much as several kilometers thick. In the oldest deposits the ice and dust are nonlayered; in the younger ones they alternate in layers, each layer some tens of meters thick.

The large quantities of dust in these deposits provide more evidence that at one time the atmosphere of Mars was denser than it is today; the present atmosphere does not seem capable of transporting such quantities of material to the poles. In fact, the most recent process has been one of partial erosion of the deposits by the wind, moving debris away from the polar deposits to form a mantle blanketing much of the surface at high latitudes. The layered deposits imply that the past climate of Mars not only was different but also may have been subject to periodic changes.

On the basis of the photographs from *Mariner 9* a number of hypotheses were proposed for the origin of the polar deposits. Some investigators suggested that the deposits of ice and dust accumulated mostly during an early period when the Martian atmosphere was denser. Others suggested that the polar deposits have been accumulating throughout Mars's history and that their accumulation has been modulated by the amount of solar radiation received by the planet, particularly at the poles. Calculations by Carl Sagan and his colleagues at Cornell University and by William K. Hartmann of the Planetary Science Institute in Flagstaff, Ariz., indicated that the amount of heat received by Mars could have been changed by the requisite amount if the luminosity of the sun has varied. Such variations are thought to occur over a period of between a million and 100 million years.

Moreover, William R. Ward of the Jet Propulsion Laboratory showed that the tilt of Mars's axis of rotation changes considerably with time because of the pull of the sun on the planet's equatorial bulge. Currently the axis of Mars is tilted from the vertical with respect to the plane of its revolution around the sun by about 25 degrees. Over a period of between 100,000 and a million years, however, the tilt of the axis changes from a minimum of 15 degrees to a maximum of 35. To add an even greater degree of complexity, more recent calculations by Ward, Joseph A. Burns of Cornell and O. Brian Toon of the Ames Research Center show that the tilt of the axis could have occasionally reached 45 degrees during an earlier period, before the formation of the great volcanic bulge Tharsis.

A radiometer aboard the early *Mariner 7* spacecraft revealed that the seasonal ice caps on Mars were composed

FROST ON MARS was photographed by *Viking Lander 2* near local noon on September 13 of last year, which was late winter in the Martian northern hemisphere. The lander is located in Utopia Planitia at a latitude of 48 degrees. The view is to the southwest over the vehicle's deck. The frost is the white patches at the base of the rocks. It is probably a remnant of a thin deposit of water ice, perhaps mixed with a clathrate in which carbon dioxide ice is caged inside water ice. The frost had slowly accumulated earlier in the winter as water was transferred to the northern hemisphere from the southern hemisphere. Analysis of the color of the surface indicates that the frost had been on the ground for at least 100 days before this photograph was made.

of frozen carbon dioxide. It seemed likely that the residual ice cap that remained during the summer was also composed of carbon dioxide ice. If the permanent ice cap consisted of carbon dioxide ice, a relatively small increase in the atmospheric temperature should increase the pressure of the Martian atmosphere from its present value of some two to 10 millibars to somewhere between 30 millibars and one bar: the atmospheric pressure at sea level on the earth. The estimate of 30 millibars, calculated by Bruce C. Murray and his colleagues at the California Institute of Technology, was based on the amount of solar radiation received by the ice caps when Mars's spin axis was at its maximum tilt of 35 degrees, and on the assumption that the atmosphere transported a minimum amount of heat to the poles. The estimate of one bar, calculated by Sagan and his co-workers, was an upper limit resulting from a runaway greenhouse effect, whereby an initial increase in temperature leads to an increase in atmospheric pressure, which allows more heat to be transported to the poles, which in turn allows even more carbon dioxide ice to evaporate.

"BIG JOE" is the name given the large dark boulder to the right of the center of this photograph made by *Viking Lander 1* at local noon on August 23, 1976. The boulder is some two meters across and is about 10 meters away from the lander; it is one of a number of boulders that can be seen lying among drifts of fine-grained material. It is likely that Big Joe was once covered by drifted material, some of which still seems to be clinging to the top of it as a yellowish brown deposit. The picture was made in the summer of the northern hemisphere, a season noted for its lack of dust storms, yet the yellowish brown color of the sky is due to suspended particles of dust. Apparently some dust is present in the Martian atmosphere most of the time. The profile of an impact crater some 300 meters from the lander can be seen on the horizon in the middle of the picture. The numbers bordering the picture aid in the interpretation of the photograph. The numbers in the vertical column at the far left show the elevation in degrees above and below the camera's horizontal plane; the numbers in the top two horizontal rows show azimuth in degrees respectively in the coordinate system of the camera and in that of the lander. Due north is at an azimuth of 130 degrees. The other numbers on photograph refer to the scan lines by which the camera built up the image.

In warmer periods, or during periods when the poles of Mars point more toward the sun than they do at present, the density of the Martian atmosphere would have been higher and more dusty material would have been eroded at low latitudes and transported to the poles. As the atmosphere cooled over the poles and condensed to form the ice cap, dust suspended in it would have been deposited along with the condensates. In cooler periods, however, so much of the atmosphere would have condensed over the poles that at lower latitudes its gaseous remnant would have become quite thin and would have transported far less dust to the poles.

The Viking Orbiter Observations

The Viking orbiters, which began their coverage of Mars early in the summer of 1976, photographed the planet with much greater clarity than *Mariner 9* did. The difference was due largely to the fact that *Mariner 9* approached Mars when the planet was still enveloped in a dust storm. Moreover, after the storm enough dust remained suspended in the atmosphere to significantly obscure the surface of Mars for several terrestrial months. The Viking orbiters began making images of Mars when the atmosphere was relatively free of dust. Moreover, although the Viking orbiter cameras were Vidicon systems similar to those carried on *Mariner 9,* they had considerably better resolution.

The Viking orbiter photographs show that much of the surface of Mars retains crisp topographic detail: lava flows, wrinkle ridges and crater ejecta stand out in sharp relief. In addition the photographs show numerous lava flows in even the most primitive cratered terrain. From these last features Michael H. Carr of the U.S. Geological Survey and his colleagues suggest that early in Mars's history even much of the ancient cratered terrain in the southern hemisphere was flooded with lava. Such volcanic leveling would explain why the highly cratered terrain on Mars is relatively smooth compared with the mountainous highlands of the moon.

The fact that the most ancient features on Mars are still sharply defined also indicates that during the planet's history there has been relatively little breakdown of rock and redistribution of debris. The only clear evidence for wind erosion on a large scale is found in regions composed of older sedimentary deposits, such as those near the poles. It is likely that the polar deposits are only partly consolidated and are therefore easily eroded by the wind.

The Viking orbiters have shown that the northern plains at high latitudes are more than simple mantles of debris. They are a complex of lava flows and deposits of windblown dust that have themselves been partially stripped by

FLOW OF DEBRIS surrounds the 25-kilometer crater Arandas, photographed by *Viking Orbiter 1* on July 22, 1976. Although the rim of the crater is clearly defined, the material around it appears to have flowed along the ground instead of being ejected as the meteorite struck the surface. At the top left the flow may have been deflected around a small crater. The radial grooves on the surface of the flow may have been eroded into it during the last stages of the impact process. Flow was probably created when heat from the meteorite's impact melted ice below the surface, and water and steam transported material away from crater in a coherent flow.

the wind. Closer to the north pole the orbiter cameras found large fields of dunes girdling the residual ice cap; the dunes seem to be composed of particles the size of grains of sand. Sand-sized particles, which on Mars probably range from .1 millimeter to several millimeters, are too large and heavy to remain suspended in the rarefied atmosphere and be carried away, but they are sufficiently small and light to be rolled and lifted for short distances by the wind. Dust particles, however, which on Mars are smaller than about .1 millimeter, are sufficiently small and light to be carried away in suspension.

The bulk of both the debris on the northern plains and the dunes closer to the poles apparently consists of material eroded from the polar deposits themselves. When the deposits were stripped by the wind, the embedded dust was carried away and any sand lagged behind and accumulated in dunes close to the poles. A puzzle here is that although the polar deposits contain dust, they seem to be yielding both dust and sand. The answer may be that the sand-sized particles are actually dust-sized particles that have been cemented together by oxides, salts or perhaps even ice.

Interpretation of the data from the radiometers on the Viking orbiters by

Hugh H. Kieffer of the University of California at Los Angeles and his associates showed that during the summer in the northern hemisphere of Mars the temperature over the residual north-polar ice cap was some 205 degrees Kelvin (68 degrees below zero Celsius). This result was striking. Since the atmospheric pressure at the surface of Mars is only about six millibars, the temperature would have to be less than 148 degrees K. to maintain a permanent cap of carbon dioxide ice. Even if the ice in the ice cap were a clathrate compound in which carbon dioxide ice was caged inside water ice, it could not exist at a temperature higher than about 155 degrees K. The only condensate that can remain stable at 205 degrees K. is water ice.

Further support for the view that the residual polar cap consists only of water ice comes from an analysis of data from the Viking orbiter spectrometers made by Crofton B. Farmer and his associates at the Jet Propulsion Laboratory. They find that during the summer at northern latitudes the amount of water vapor in the atmosphere is such that the temperatures must be higher than 200 degrees K.

Observations of the residual south-polar cap during the summer in the southern hemisphere are difficult to interpret because of the effect of a global dust cloud that may have modulated atmospheric temperatures. The residual southern ice cap is probably water ice too, although measurements of its temperature are ambiguous, and there is a slim chance that its major constituent is carbon dioxide ice.

Since it seems that there is water ice at both poles, hypotheses involving water as an active agent in Mars's past have gained measurably more acceptance. Moreover, the Viking orbiter photographs have revealed that the channels on Mars are even more abundant than was indicated by the *Mariner 9* photographs and that the channels extend to smaller sizes and form a much more integrated drainage system than had been previously perceived. The runoff of rain from an ancient dense atmosphere may have carved some of the treelike networks of channels. Other channels may

have been formed when underground ice was melted by heat released in volcanic activity. Then the channels would have been created when the ice melted, the ground above it slumped and the water flowed away. In this case the formation of the channels would seem to be directly linked to the thermal history of the planet.

Intriguing evidence for the presence of water ice in the crust and regolith of Mars has also been obtained from examining pictures of the peculiar terraces, ramparts and lobes characteristic of the ejecta of many of the large craters. On the moon the ejecta from impact craters appear to have been blocks of material that were hurled outward by the original impact and fell back to the surface, excavating myriads of secondary impact craters. On Mars the ejecta deposits surround the impact craters al-

most like a solidified flow. The probable explanation is that the heat of the impact melted and vaporized water ice trapped in the crust, and the liquid water and steam transported the ejecta away from the crater in a coherent, ground-hugging flow of debris. In some photographs one can even see where the flow was diverted around obstacles in its path.

Although water is now being recognized as an important agent in Mars's past, the discovery that the polar regions are probably dominated by water ice instead of carbon dioxide ice places severe constraints on the intensity of past climatic fluctuations. If a water-ice cap on Mars received more sunlight, it would begin to evaporate, subliming directly from the solid ice to a vapor. In order for the vapor pressure of water on Mars to reach the pressure required for liquid

CHRYSE PLANITIA, the site of *Viking Lander 1,* is a rolling plain littered with blocks. Just visible to the right of the center are several areas of exposed bedrock. Below them small linear deposits or streaks of sediment can be seen extending away from the rocks. The streaks extend roughly in a north-to-south direction. Toward the upper left is a large field of drifted material where deposits accumulated during a period when winds were also blowing from north to south. Because of the panoramic geometry of the image it seems that the streaks change direction with azimuth. The center of the mosaic is pointing toward the southeast. The entire mosaic covers 160 degrees in azi-

water to exist the temperature would have to be raised by at least 70 degrees K. Such a dramatic rise in temperature would be most unlikely to occur even if the luminosity of the sun varied by the maximum amount allowed by theory or if the inclination of the planet's axis periodically changed to the maximum extent. Indeed, the fact that the most ancient surfaces of Mars are so well preserved is consistent with the hypothesis that the atmospheric conditions on Mars have not fluctuated greatly over most of the planet's history. It seems likely that the bulk of the polar deposits formed very early and have since been eroded by the wind. The exact time they formed and the reason for their formation, along with the history of any early, dense atmosphere that may have been supported by greenhouse effects, however, remain a mystery.

Viking Lander 1 touched down on the western slopes of the region named Chryse Planitia (22.5 degrees north latitude, 47.8 degrees west longitude). From orbit the landing site looks much like the surface of a lunar mare, or "sea": it is a smooth volcanic plain sparsely cratered and lined with a series of wrinkle ridges. The walls of the craters seem nearly pristine and the wrinkle ridges are also well preserved. The amount of erosion must be very slight or confined to a scale on the order of meters for the morphology of the ejecta deposits and the ridges to be so little changed.

The landing site is some 130 kilometers east of Lunae Planum, one of the most heavily cratered plains on Mars. Lunae Planum is about a kilometer higher in elevation than Chryse Planitia, and the boundary between the two re-

gions is marked by an irregular escarpment. A number of large channels course through Lunae Planum, emerge from the escarpment and extend eastward across Chryse Planitia toward the landing site. The channels were most likely created by ground water from ice trapped within Lunae Planum. At some time in the past volcanism and geothermal heating melted some of the underground ice and the water escaped to the surface to create one or more torrential floods that cut into the Lunae Planum deposits and poured onto Chryse Planitia. The flow of water from Lunae Planum breached several craters and wrinkle ridges to the west of the landing site.

The Surface of Chryse Planitia

To judge from correlations between pictures from *Viking Lander 1* and pic-

muth. The local topography is such that the distance of the horizon ranges from several tens of meters at the left to several kilometers at the right. The terrain at the bottom is only a couple of meters from the camera. Scale is provided by the surface sampler assembly pin, the bright cylindrical object lying in the soil in the bottom middle por-

tion of the mosaic; it is 10 centimeters long. The lander's footpad is at the bottom right, and other parts of the vehicle can be seen to the left. Under the spacecraft the retrorocket exhaust blew loose material away, exposing a fractured, crusted type of soil called duricrust; a few chunks of such soil are visible to the left of the assembly pin.

tures from the Viking orbiters, the lander is sitting on the flank of a wrinkle ridge. From the lander the site strikingly resembles many rocky deserts on the earth, particularly those with exposures of volcanic rock. The gently rolling landscape is yellowish brown in color, strewn with rocks and dotted with drifts of fine-grained material. Within 30 meters of the spacecraft several outcrops of bedrock can be discerned. No positive evidence of the flood from Lunae Planum is visible on the photographs from the lander: scoured features, channels or fluvial deposits cannot be detected. Apparently the flood either did not reach the landing site or had largely dissipated by the time it passed over it. Alternatively the surface may have been so modified since the flood that any fluvial features are no longer discernible.

The resemblance of the landing site to rocky deserts on the earth was somewhat surprising on the basis of what

most investigators expected from the moonlike orbital pictures of Chryse Planitia. From orbit the dominant features of the region are craters. From the ground there are only a few obvious craters to be seen in the immediate vicinity of the lander. Based on the population of large craters visible from orbit, Edward A. Guinness, Jr., of Washington University calculated that if Mars were like the moon, some 35 craters with diameters ranging between 25 and 50 meters should be in the lander's field of view.

A relative deficiency of craters on Mars smaller than 50 meters across was actually predicted in 1970 by Donald E. Gault and Barrett S. Baldwin, Jr., of the Ames Research Center. They calculated that even though the atmosphere of Mars is thin, it is dense enough to ablate and break up small incoming meteoroids before they reach the surface. As a result the Martian surface is not subject

ed to the repetitive high-velocity impact of small objects and the consequent "gardening" of the top few meters of the surface. The only substantial population of craters with diameters of less than 50 meters should be secondary craters produced by the impact of debris thrown out during the formation of a large crater (one that is tens of kilometers in diameter). In contrast, the surface of the moon shows a continuous spectrum of crater sizes, and the impact of small objects over millions of years has created a dusty soil. On Mars large impacts would fracture the surface and strew over it a discontinuous layer of relatively large blocks, just as at the landing site. The soil on Mars must have been created by other processes.

In addition to the rocks and outcrops visible on the ground at Chryse Planitia, fine-grained material is abundant in the form of streaks on the lee side of most of the rocks; the streaks are several centi

UTOPIA PLANITIA, the site of *Viking Lander 2*, is superficially like Chryse Planitia in that the terrain is littered with blocks. The terrain is remarkably flat, however, and the horizon is several kilometers **away. Scale is provided by the largest block in the middle of the image, which is 2.75 meters from the spacecraft and is 35 centimeters wide. No bedrock seems to be exposed. A prominent troughlike depression**

meters in depth and range from 10 centimeters to a meter in length. To the northeast of the lander there is a complex of drifts in a field of boulders; these drifts probably were formed when windblown material was trapped among boulders that were sufficiently large to reduce the velocity of the wind locally. Several of the large drifts appear to be layered or laminated. On the earth layers are usually not seen in actively growing or moving drifts; they are visible only when the drifts are stabilized by vegetation or cementation and are being excavated by wind erosion. It appears that on Mars the drifts observed were deposited some time ago, that they were partially lithified (formed into sedimentary rock) and that they have recently been eroded.

On the average the long axes of the windblown streaks behind the rocks point almost due south. Photographs made by *Mariner 9* of regions just to the

north of the landing site show that large streaks extending from craters also point roughly south. Both sets of streaks seem to indicate that the prevailing winds near the surface had blown from north to south during the period when the streaks were formed. Moreover, the pattern of exposed layers on the drifts indicates that when the drifts were deposited, the wind direction was also north to south.

Another feature at the landing site was revealed by the landing itself. Close to the spacecraft the retrorocket exhaust blew away loose material, exposing a crust of soil fractured into a polygonal pattern. Such soil, known as duricrust, is similar in appearance to the deposits called caliche in the U.S. Southwest and Mexico. On the earth duricrust is formed when dilute solutions of salt migrate up through the soil; the water evaporates from the solutions, and the salts and other substances collect just

below the surface. The same process probably has operated on Mars. It is not known whether on Mars the water comes from relatively large pores below the surface or from thin films of water between grains of material. It probably comes from the thin films, because variations in the water-vapor content of the lower atmosphere suggest that water is regularly cycled between the surface and the atmosphere.

Direct evidence that salts are present in the Martian duricrust was obtained by sampling the soil in front of the lander and analyzing its chemistry with the X-ray-fluorescence spectrometer on the lander. Priestley Toulmin III of the U.S. Geological Survey and the members of his X-ray team found that the amount of sulfur, a likely candidate for being bound in salt minerals, was somewhat greater in clods of soil than it was in loose soil. The clods are abundant around the lander, and they probably

about a meter wide and 10 centimeters deep cuts horizontally through the middle of the picture. Several drifts of material occupy its bottom. Linear deposits or streaks extend downwind of the rocks at the far left. The streaks have approximately the same azimuth as those behind the rocks at Chryse Planitia. The bright plateau on the horizon to the right is approximately in direction of the large crater Mie.

are chunks of duricrust broken up by the landing and also by the natural wind.

The Surface of Utopia Planitia

The search for the landing site for *Viking Lander 2* was based partly on the desire of investigators to have the second lander in a region distinctly different from that of *Viking Lander 1* but still smooth enough for a successful landing. The overriding concern, however, was to find a location with a high water-vapor content in order to maximize the probability of finding evidence of life. The region chosen was the surface of Utopia Planitia, which is part of the mantle of debris on the vast plains of the northern hemisphere. Observations from orbit showed that the surface is cut by fractures that break the terrain into polygonal forms kilometers on a side. The landing site (48 degrees north latitude, 225.6 degrees west longitude) is 200 kilometers southwest of the large crater Mie, 100 kilometers across.

Soon after *Viking Lander 2* had settled on the surface it was apparent that from the ground the site of *Viking Lander 2* was superficially similar to the site of *Viking Lander 1:* the surface is a rock-strewn plain with duricrust. That, however, is where the resemblance ends. *Viking Lander 2* came down on a flat plain, where most of the topographic relief is created by troughlike depressions that divide the terrain into polygons. No outcrops of bedrock can be discerned. Instead boulders and cobbles either are partly embedded in a matrix of fine-grained material or are resting on top of it. Although other explanations for the appearance of the landing site are possible, it is likely that *Viking Lander 2* came to rest on a lobe of the debris that flowed out from the crater Mie. In flows of debris on the earth large rocks are commonly carried on top of the flow, leaving a field of boulders partly embedded in finer-grained material.

The troughlike depressions visible from *Viking Lander 2* are approximately a meter wide and 10 centimeters deep, and the edges of the troughs are slightly turned up. The troughs visible from the lander are much smaller than the troughs seen from orbit, but it is likely that the features have a similar origin. The scale and the form of the troughs seen from the lander make them a good physical analogue of the "patterned ground" that is found in cold regions on the earth. On the earth patterned ground forms in ice-saturated soil where low temperatures cause the ground to contract and fracture; the fractures usually cut the ground in a polygonal pattern. In the spring the frozen soil thaws and the fractures fill with water. In the fall the water freezes, and in the following winter the ice, which is weaker than the frozen soil, fractures in the same pattern. Repeated cycles of the process create a terrain that is cut by troughs in a polygonal pattern, with the troughs occupied by wedges of ice.

One problem with the proposal that this process is currently operating on Mars is that the temperatures at Utopia Planitia are always below the freezing point of water. Benton C. Clark of the Martin Marietta Corporation has estimated that even if the ice contained salts, the freezing point of the solution would be depressed only 10 or 20 degrees K. below the melting point of pure water. At the temperatures on Utopia Planitia the ice would still never melt.

Patterned ground can be created by one other process: the desiccation of clays. Clays, which are the leading candidate for the major constituent of the Martian soils sampled at the landing sites, expand and contract by as much as 20 percent as they absorb water and then lose it. If the ground on which the *Viking Lander 2* is resting was once saturated with water and then dried out, cracks may have formed, creating the troughs in a polygonal pattern.

Small streaks can be seen extending downwind from some of the rocks at the site of *Viking Lander 2*, just as they do at the site of *Viking Lander 1*. In addition several small drifts occupy the floor of a large trough in front of the lander. Again the drifts indicate that the prevailing wind direction is roughly from north to south; thus Chryse Planitia and Utopia Planitia may have been subject to the same wind system. Since the sites are halfway around the planet from each other, such a wind system would have to have been global. Most likely both sets of wind streaks and drifts formed during dust storms at Mars's closest approach to the sun, when the flow of the atmosphere could have been in a north-to-south direction in the northern hemisphere. The Viking lander meteorology stations, which monitored winds during the 1977 dust storms, did not, however, show any predominance of a north-to-south flow.

The Martian Soil

The X-ray-fluorescence spectrometer on each lander analyzed soil samples and determined the abundance of a number of elements with atomic numbers higher than that of sodium (atomic number 11). Toulmin and his colleagues have examined the data and have shown that the overall composition of the soil at one site is much the same as that at the other. The composition is different from that of any single known mineral or type of rock, indicating that the soil is probably a complex mixture of materials.

The surface of Mars seems to be composed of soil derived from mafic igneous rocks, that is, rocks that have crystallized from a melt that was rich in mag-

TWO DRIFTS OF FINE-GRAINED MATERIAL are 15 meters from *Viking Lander 1*. The drift at the right is composed of darker material than the drift at the left. Spectra of the dark drift indicate that it is the only example of such material at either landing site. All the other soil areas seen from the landers are quite similar. They are probably the same as the soils that dominate the bright areas of Mars.

nesium and iron. Compared with rocks on the earth in general, they are rich in magnesium, iron and calcium and poor in potassium, silicon and aluminum. Such abundances are compatible with the kind of materials one would expect from a partial melting of the Martian mantle: the deep layer below the crust.

It is likely that the soil analyzed by the Viking X-ray spectrometers is a mixture of iron-rich clay minerals, iron hydroxides, sulfate minerals and carbonate minerals. Such a deduction is consistent with the results from the combined gas-chromatograph and mass-spectrometer experiments on each lander, which found that when samples of soil were heated, water vapor and carbon dioxide were released. The soil contains about 1 percent water by weight, some of which is probably in hydrated minerals.

On the earth mafic materials are chemically altered by water and give rise to iron-rich clayey soils. The same kind of process could have operated in the Martian past, when liquid water was prevalent. Some of the soil could also have formed when hot magmas penetrated the ice-laden crust and regolith, erupting explosively to form clay tuffs. In addition, if a sufficient quantity of water was available, the heat from impacts may have been sufficient to alter volcanic rock materials into clays.

An intriguing alternative hypothesis has been proposed by Robert L. Huguenin of the University of Massachusetts. He suggests that the soil has largely been produced by oxidation stimulated by ultraviolet irradiation of the rocks. Ultraviolet radiation from the sun is not absorbed by the Martian atmosphere because the atmosphere does not have an ozone layer. In the presence of small amounts of water vapor such radiation is capable of breaking down aluminosilicate minerals by causing ions such as those of iron to migrate toward the surface, thus disrupting the minerals' crystal structure. The extent to which the soil might have been created by such a process is not known.

Two magnets were mounted on the hoe that was part of the soil-acquisition scoop on each lander. One of the magnets was mounted flush with the hoe's surface; the other was embedded in the metal so that its effective strength was a twelfth that of the first. At both sites equal amounts of material clung to both magnets after the hoe had been inserted into the soil. According to Robert B. Hargraves of Princeton University and David W. Collinson of the University of Newcastle upon Tyne, for equal aggregates of material to cling to both magnets between 3 and 7 percent of the material by weight must be magnetic. If the soil had a lower concentration of magnetic material, the aggregates would not cling to the weaker magnet. Hence magnetic material is present in significant amounts in the Martian soil.

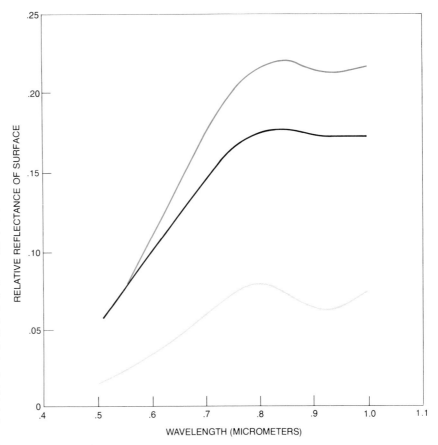

REFLECTANCE SPECTRA are shown for the bright drift and the dark drift at the site of *Viking Lander 1* and also for a trench dug there by the surface sampler. The spectra were produced at the Langley Research Center of the National Aeronautics and Space Administration with a specialized technique devised by Friedrich O. Huck and Stephen K. Park. The spectra show the fraction of sunlight reflected at wavelengths between .5 micrometer (in the blue) and 1.0 micrometer (in the near infrared). The shapes of the spectra for the bright drift (*dark color*) and for the trench (*light color*) are similar. Both show an absorption band at .93 micrometer, which implies that they have a similar composition. The only major difference between them is in the intensity of the light reflected. (The difference is actually to be expected because the soil in the trench was disturbed when the trench was dug, thereby increasing the soil's microtopography, increasing its degree of scattering and shadowing and decreasing its reflectance.) The reflectance spectrum of the dark drift (*black*), however, has a different shape. The absorption band at .93 micrometer is gone, making the spectrum look flat, and there may be a shallow band near 1.0 micrometer. The difference between the bright drifts and the dark ones may mimic the difference between large bright and dark areas on Mars as it is seen from the earth.

The color of the material clinging to the magnets is the same as the color of the surface, which indicates that the magnetic materials are covered or stained with the same materials coloring the Martian soil. Reasonable candidates for the magnetic minerals are magnetite and maghemite, both of which are iron oxides, and metallic nickel-iron. Maghemite is yellowish brown or reddish in color, and if it is present, it may contribute to the color of Mars. The abundance of the magnetic material and its probable nature is also consistent with a source that is mafic, such as mafic basalts.

The cameras on the Viking landers made pictures of the surface and the sky in six regions of the spectrum, ranging from about .5 micrometer (in the blue) to 1.0 micrometer (in the near infrared). The color pictures were generated by first determining the spectral irradiance

of Mars in each of the regions and then computing the hue, brightness and saturation of color for the range of wavelengths to which the human eye is sensitive. The spectral irradiance of the surface of Mars is a product of the irradiance from the sun, the reflectance of the surface of Mars and the character of the light scattered by the atmosphere. The color pictures show that the surface is yellowish brown. If the effects of the color introduced by the atmosphere are removed, the color of the surface tends even more toward brown.

The reflectance spectra derived from the camera data for various soil areas at both landing sites are remarkably similar and seem to vary only as the lighting changes. The spectra are similar in appearance to spectra made of bright areas on Mars with instruments on the earth. Those areas are thought to be covered with a fine-grained, chemically altered soil. An explanation fitting the available data is that most of the soil exposures seen at the landing sites and sampled for analysis consist of a bright chemical weathering product that has been mixed on a global scale by the wind.

The one area seen at the landing sites that has a reflectance spectrum appreciably different from the spectrum of all the other areas is a drift of material some 15 meters away from *Viking Lander 1*. This drift, which is significantly darker than the surrounding soil, may have been left behind after the wind carried away brighter material that was more easily moved. The spectrum for the dark drift is similar to the spectra of the dark areas on Mars that have been made with instruments on the earth. The drift may be composed of partially altered iron-rich igneous rocks that have

been broken down into soil particles. Thus although in general the similarities between the soils at the two sites are strong, differences do exist that may mimic on a small scale the differences between the classic bright and dark areas of Mars seen from the earth.

Most of the rocks and bedrock outcrops in the color pictures made by the landers appear to be darker than the soil on which they rest, but when they are viewed under comparable illumination, most of them are actually brighter. In the majority of the photographs the angle between the sun, the surface and the camera is larger for the sides of the rocks than it is for the soil on the ground, and hence the brightness of the rocks appears lower. The rocks and soil have similar spectral shapes, indicating that the rocks are covered with a smooth

stain or dusting of material similar in composition to the soil. The general appearance of most of the rocks suggests that they are volcanic in origin, although some have been eroded significantly by the wind. Most of the rocks at the site of *Viking Lander 2* and some of those at the site of *Viking Lander 1* are pitted like the ones that form from gas-rich lavas on the earth. Gases dissolved in the lava because of the confining pressures at great depths are liberated when the lava reaches the surface and form pockets of gas. When the molten rock cools and hardens, the pockets are preserved as bubbles or vesicles in the rock. Unfortunately distinguishing between pits produced by this process and those produced by the wind erosion of softer minerals is extremely difficult.

One of the more intriguing aspects

CLOSEUP VIEWS OF MARTIAN ROCKS illustrate the variety of rocks on Mars. The top two photographs show rocks from Chryse Planitia; the bottom two photographs show rocks from Utopia Planitia. The rock at the top left is about 25 centimeters long; its pitted or mottled nature is reminiscent of igneous rocks on the earth that have been abraded by the wind. Duricrust exposed by the retrorocket exhaust can be seen to the lower left. The rock at the top right appears to be a breccialike volcanic rock about 20 centimeters across. A moat scoured by the wind can be seen on one side of the rock; a raised deposit of windblown material extends from the other side. The elongated pits in the soil were made when clods hurled out by the landing hit the ground. The rock at the bottom left is a rectangular block about 40 centimeters wide. The angular form of the rock indicates that its appearance is dominated by fracture planes at right angles to one another. Drifted material is visible in the background, along with a conical rock that may have been shaped by wind erosion. The peanut-shaped rock at the bottom right may be a piece of lava some 30 centimeters wide. The pits in it may have been created when gas dissolved in the lava formed bubbles or vesicles and then escaped, or they may have been pockets of softer minerals preferentially eroded by the wind. The landing sites display a variety of rock forms, some eroded, some appearing pristine. Most seem to be of volcanic origin.

of both landing sites, and probably of Mars in general, is the fact that there seems to be a marked deficiency of sand compared with a typical desert on the earth. The lack of sand at the landing sites has been inferred by Henry J. Moore II of the U.S. Geological Survey and his associates from photographs of the walls of the trenches that were dug by the soil samplers of both Viking landers. Their results indicate that the bulk of the small particles on Mars are smaller than .1 millimeter. Such particles would be difficult to erode from the surface, and once eroded they would probably be carried away by the wind.

The Winds of Mars

Particles the size of those of sand exist on Mars, probably in the dunes that girdle the Martian ice caps; perhaps they even make up the dark drift exposed at the site of *Viking Lander 1* and parts of the classic dark areas that cover the middle latitudes of the southern hemisphere. The Martian soil does not, however, contain the kind of sand found on the earth. Most terrestrial sand consists of quartz and feldspar minerals that have been weathered from acidic igneous and metamorphic rocks. Quartz and feldspar, which dominate acidic rocks, are resistant to chemical weathering and mechanical attack, and the two minerals are major constituents of sedimentary rocks on the earth. Mars has not differentiated to the extent that acidic rocks have been created in large quantities; the planet is probably dominated by mafic basalts. In such basalts the major minerals are olivine, pyroxene and plagioclase feldspar. Huguenin has shown that even in the cold, arid environment of Mars these minerals are rapidly weathered by ultraviolet-stimulated oxidation.

If, as is suspected, the bright soil at the landing sites is made up of clay minerals, the soil would be very fine-grained. Particles the size of grains of sand could exist, however, as aggregates of smaller particles. Such sand-sized aggregates would have a short lifetime in the Martian aeolian system. Grains bouncing along the surface travel roughly at the speed of the wind. Since Mars's atmosphere is only about a hundredth as dense as the earth's, winds of about 50 meters per second (180 kilometers per hour) would be needed to erode sand-sized particles from the Martian surface. Particles of terrestrial sand carried at that velocity would be powerfully erosive, but aggregates of clayey material would rupture on impact into harmless motes of dust. Even mineral grains such as olivine, pyroxene or feldspar would be likely to rupture if they were carried at such velocities. One consequence of this situation on Mars is that the rate of aeolian erosion is probably low compared with the rate on the earth.

The low rate of erosion probably explains why much of the Martian surface looks so crisp and pristine. The loose material can be moved around readily, but the rock erodes very slowly.

The skies at the landing sites are yellowish brown, and have remained that color over the Martian year the landers have been on Mars. This observation is somewhat unexpected. According to calculations by Pollack, the color is due mainly to particles of dust suspended in the atmosphere up to a height of 40 kilometers. It was expected that large amounts of dust would be suspended in the atmosphere after the major dust storms, which occur when the planet passes perihelion, the point on its orbit where it is closest to the sun. The lander photographs showed, however, that the skies were yellowish brown even when Mars was at aphelion, the point on its orbit where it is farthest from the sun. At that time major dust storms are rare. Dust either is often raised on Mars or is dynamically supported by atmospheric turbulence for long periods of time. Even on the dustiest days, however, if all the dust were precipitated onto the surface, the layer would be only a fraction of a millimeter thick.

The landers have gathered data during two large dust storms, both of which began in the southern hemisphere as Mars neared perihelion in 1977. The first storm began in February and the second in May. The dust cloud was quickly distributed around the planet by high-altitude winds. Each storm took several months to subside. At the landing sites in the northern hemisphere the wind only rarely reached velocities sufficiently high to disturb the soil on the ground. It seems probable that after both storms a thin layer of dust accumulated on the landers and on the surfaces around them. The layer of dust indicates that not all the material raised from the southern hemisphere is returned. Over a period of time the dust storms should denude the middle-southern latitudes of dust, exposing the bedrock and leaving behind deposits of darker, less weathered material. Perhaps such deposits are the classic Martian dark areas seen from the earth.

The latitude on Mars that is directly below the sun at perihelion is the point at which most dust storms begin. This perihelion subsolar point slowly varies with time because the axis of Mars precesses. Martian precession should cause the latitude of the perihelion subsolar point to migrate between +25 degrees and −25 degrees in latitude with a period of 50,000 years. Lawrence A. Soderblom of the U.S. Geological Survey has pointed out that if the dark areas on Mars are regions stripped of a relatively large fraction of the bright layer of mobile dust, they may migrate around the equator with the same period. In other words, some 20,000 years ago most of

the perihelion dust storms may have begun at the latitude of *Viking Lander 1*, and Chryse Planitia may have been partially stripped of its bright deposits. If that is the case, the drifts and streaks seen at the site of *Viking Lander 1* may be younger than 20,000 years.

The Future

It is clear that the Viking mission has greatly increased our knowledge of the geology of Mars. We now have a good understanding of what the surface looks like and of the kind of surface materials present. The discovery that at least one of the residual ice caps consists of water ice has significantly added to our knowledge of the intensity of the planet's climatic fluctuations. There remain important questions about the evolution of the Martian surface. The ages of the planet's various terrains are not known with much certainty. The structure and composition of the interior of Mars are still largely a mystery. Without such knowledge it is not possible to construct a unique theoretical model of how the planet formed and evolved.

Some of the questions may be answered by further analysis of the Viking data. Others will have to await future missions. One possible mission is an orbiter, much more sophisticated than the terrestrial satellite Landsat, capable of mapping detailed chemical and mineralogical characteristics of the Martian surface. Another possibility is an unmanned surface rover capable of traversing hundreds of kilometers over a period of several years, which would be able to analyze the regolith in more detail than any satellite in orbit could. A third possibility is a series of penetrator rockets launched from an orbiting platform that could embed themselves in the planet's surface at various points and provide a network of meteorological and seismological sensors. A fourth possibility is a mission that would return samples of Martian material, affording the kind of data that can be obtained only in a terrestrial laboratory. One need only consider the vast amounts of data that were obtained by analyzing the samples returned from the moon in the Apollo missions to appreciate the extent of the information that could be gained from samples returned from Mars.

The future exploration of Mars can be justified from a number of points of view. Perhaps the most important is that the earth and Mars seem to have followed evolutionary tracks that are sufficiently different for the two planets to have had unique histories. Their evolutionary tracks have been sufficiently similar, however, to provide for meaningful comparisons of data on their atmosphere, surface and interior. There can be little doubt that understanding the history of Mars will increase our understanding of the earth.

6

Jupiter and Saturn

by Andrew P. Ingersoll
December, 1981

Competing models seek to describe the sun's two giant companions. In one model the winds are confined to a thin layer at the surface; in another the winds extend through the fluid depths of each planet

There were many exciting moments during the encounters of the Voyager spacecraft with Jupiter in 1979 and with Saturn in 1980 and 1981, but to me the most memorable ones came early in the first encounter, when we viewed the time-lapse images of Jupiter's swirling clouds made by *Voyager 1* on its long approach to the planet. The sequence compressed a 30-day history of the Jovian weather into a one-minute motion picture. As a scientist who studies the atmospheres of the planets, I was familiar with Jupiter's brown belts and white zones: the colored cloud bands some thousands of kilometers wide that circle the planet at constant latitude. And I had come to accept the hypothesis that Jupiter's Great Red Spot is a storm as large as the earth that has lasted for centuries.

I was not prepared, however, for the intricate motions revealed in the time-lapse sequence. The sequence was synchronized to the rotation of Jupiter, so that the Great Red Spot appeared to be stationary, with the adjacent atmosphere swirling around it. Bright, small-scale features appeared and then were torn apart, all in a few days. Small spots encountering the Red Spot from the east seemed to be drawn into its counterclockwise, rotatory flow, to circle it in one or two weeks and then to divide, a part of each one merging with the spot and the other part returning eastward. Elsewhere spots were forming, merging and dividing every few days. On scales less than 1,000 kilometers everything seemed to be chaos. How the larger structures could endure and retain their distinct colorations in such a well-mixed atmosphere was more of a mystery than it had been before the spacecraft arrived.

It is still a mystery today. Nevertheless, an analysis of the images made by *Voyager 1* and *Voyager 2* reveals order in the chaos. Indeed, it shows that many aspects of the atmospheric circulation on Jupiter and Saturn resemble patterns in the atmosphere and oceans of the earth. A number of theoretical models and laboratory experiments are now in-

voked to explain the circulation. The models differ in their assumptions about a basic unanswered question. Are the motions in the atmosphere of Jupiter and Saturn confined to a thin sunlit layer 100 or 200 kilometers thick where clouds form and the atmospheric pressure is only a few times the sea-level pressure on the earth? Or do the motions extend tens of thousands of kilometers down to the top of a metallic hydrogen zone where the pressure is three million times the earth's sea-level pressure?

In considering these questions I shall focus on Jupiter and Saturn themselves, to the exclusion of their moons and rings. I shall begin with bulk planetary properties: the mass, the density and the composition of Jupiter and Saturn. This leads to a discussion of the internal structure of each planet and of the origin of the internal heat each is radiating into space. Next I shall describe the structure and chemistry of the atmosphere of Jupiter and Saturn. Alternative explanations of the dynamics of the atmosphere I shall take up last. Here the challenge is to find out why features of atmospheric flow can persist for decades or centuries on Jupiter and Saturn and only for days or weeks on the earth.

Mass, Density and Internal Structure

No theory is available that accounts precisely for the mass of the sun's two principal companions. The mass of Jupiter is 318 times the mass of the earth, or about a thousandth the mass of the sun. The mass of Saturn is 95 times the mass of the earth. Presumably these values reflect the amount of matter left in orbit around the sun soon after it formed. Moreover, no theory is available that accounts precisely for the relative abundances of the various chemical elements constituting Jupiter and Saturn. The abundances do, however, resemble those in the sun. Specifically, Jupiter and Saturn are the only planets in the solar system that consist mostly of hydrogen and helium. No other substances could give Jupiter its bulk density of only 1.33 grams per cubic centime-

ter at the pressures and temperatures that characterize the planet. The density of Saturn is even less—.69 gram per cubic centimeter—because Saturn's smaller mass entails a lesser degree of gravitational self-compression.

Mercury, Venus, the earth and Mars are made of heavier stuff: their densities are from 3.9 to 5.5 grams per cubic centimeter, or several times the density of Saturn. In general they are made of rocks: the most abundant metals and their oxides. The densities of Uranus and Neptune are twice as great as Saturn's, although their self-compression is less. Their most likely constituent elements are therefore oxygen, carbon and nitrogen: the third, fourth and fifth most abundant elements in the solar system. (The first two are hydrogen and helium.) At the temperatures that characterize the outer solar system oxygen, carbon and nitrogen should combine with the available hydrogen to form water, methane (CH_4) and ammonia (NH_3). On the surface of Uranus and Neptune these compounds are ices; in the interior they are liquids. The enrichment of ices with respect to hydrogen and helium on the planets beyond Saturn is hard to explain. The depletion of gases and ices in the inner solar system probably reflects the high temperatures close to the sun in the early life of the solar system.

Given the mass and composition of a planet such as Jupiter or Saturn one may seek to infer its internal structure. The size and density of the planet adjust themselves so that the outward pressure of the compressed material exactly balances the inward pull of gravity at any given place in the planet's interior. The result is a state of hydrostatic equilibrium. If the planet is rotating, a further force enters the balance. It is the outward centrifugal force that results from the planet's rotation. The outward force on a rotating mass is proportional to the square of the distance of the mass from the axis of rotation. Hence a rotating planet is flattened: its polar radius is less than its equatorial radius. The degree of flattening will depend on the internal distribution of mass. For example,

EDDY CURRENTS in the atmosphere of Jupiter and Saturn are deviations in patterns of flow that otherwise consist of sustained, alternating ribbons of eastward and westward wind. The northern mid-latitudes of Jupiter (*top*) were photographed in exaggerated color by the spacecraft *Voyager 1* on March 2, 1979. The orange ribbon cutting diagonally across the bottom right corner represents a steady eastward wind whose speed is some 130 meters per second. The sinu-ous lines toward the top represent eddies whose speeds are about 30 meters per second with respect to the steadier currents. The northern mid-latitudes of Saturn (*bottom*) were photographed in exaggerated color by *Voyager 2* on August 19 of this year. The sinuous line inside the light blue ribbon is a pattern moving eastward at 150 meters per second. The dark oval and the puffy white features below it are eddies drifting westward at speeds as high as 20 meters per second.

two planets of the same mass and rate of rotation will differ in the degree to which they are flattened if matter is concentrated near the center in one and farther from the center in the other. The latter will be the more flattened of the two. Clearly the degree of flattening is a sensitive probe of a rotating planet's internal structure.

Both Jupiter and Saturn have a rotational period of about 10 hours. Moreover, both planets are somewhat flattened. Jupiter's equatorial radius is 6.5 percent greater than its polar radius; Saturn's is 9.6 percent greater. Measurements of the planets' gravitational field imply a corresponding degree of concentration of mass toward the equatorial plane. An incorporation of the measurements into models of the internal structure of Jupiter and Saturn have led William B. Hubbard, Jr., of the University of Arizona and V. Zharkov and V. Trubitsyn of the Institute of the Physics of the Earth in Moscow to a further conclusion. Both Jupiter and Saturn have a dense core that cannot consist of compressed hydrogen and helium. The pressure inside each planet is simply not great enough to produce the required central densities from a mixture of those two elements. Apparently Jupiter has a core of rock and ice that constitutes about 4 percent of its mass, and Saturn has a similar core that constitutes about 25 percent of its mass. Each core may be the "seed" on which the rest of the planet condensed from gases when the solar system formed. Or perhaps the cores formed later as the result of a redistribution of matter inside the planets.

Knowledge of the internal structure of Jupiter and Saturn also comes from the quantum-mechanical description of how atoms and molecules behave as they are compressed. According to the exclusion principle of modern physics, the electrons bound to protons in a compressed assemblage of hydrogen molecules can occupy the same shrinking volume only by climbing to higher levels of energy. At a certain compression (and thus a certain amount of energy) they are no longer bound to individual protons but become free to wander in an electrically neutral mixture of protons and electrons. The hydrogen then becomes a metal. Calculations made by Edwin E. Salpeter of Cornell University and David J. Stevenson, who is now at the California Institute of Technology, show that the transition from molecular hydrogen to metallic hydrogen comes at nearly the same critical pressure (three million earth atmospheres) on both Jupiter and Saturn. Since Jupiter is more massive than Saturn, the critical pressure is attained closer to the surface. On Jupiter the distance from the center of the planet to the metallic-molecular transition is in the range of .75 to .80 times the distance from the center to the surface. On Saturn it is .45 to .50.

Sources of Internal Heat

Measurements made by instruments on the Voyager spacecraft, and also the Pioneer spacecraft that reached Jupiter as early as 1973, imply that the power Jupiter gives off (as infrared radiation) is 1.5 to 2.0 times the amount it absorbs (as sunlight). The power Saturn gives off is 2.0 to 3.0 times the amount it absorbs. Hence both Jupiter and Saturn have internal sources of heat. Yet neither planet is massive enough for gravitational self-compression to have initiated nuclear fusion. In short, neither planet is a star. Instead their internal heat must represent the conversion of the gravitational potential energy that became available as each planet contracted from a cloud of gas beginning some 4.6 billion years ago. James B. Pollack and his colleagues at the Ames Research Center of the National Aeronautics and Space Administration have developed models of the history of the giant planets and conclude that the interior of Jupiter and Saturn is still hot today.

How hot? The answer follows from the thermodynamic law that heat flows from warmer places to colder. Specifically, a mixture of hydrogen and helium in Jupiter or Saturn (even a metallic mixture) cannot conduct heat away from the center of the planet unless the rate of temperature decrease with distance from the center is significant. The rate of decrease is limited, however, by convection: the overturning of a fluid in which warm parcels of the fluid rise and cooler parcels sink.

Briefly, convection mixes a fluid until its temperature decrease with altitude matches the adiabatic lapse rate: the rate at which a rising fluid parcel will cool when no heat is exchanged with its surroundings. Instead of losing energy by heat flow the parcel loses energy by pushing on its surroundings as it expands. On this basis it can be shown that the temperature gradient in Jupiter and Saturn's interior is close to adiabatic and that the central temperatures are in a range of 20,000 to 30,000 degrees Kelvin. At such temperatures a mixture of hydrogen and helium does not solidify. Thus the metallic hydrogen inside Jupiter and Saturn is liquid. At an intermediate level, where the pressure is three million earth atmospheres, the metallic liquid abruptly gives way to a molecular liquid. At still higher levels the molecular liquid gradually gives way to a molecular gas: the atmosphere of Jupiter and Saturn.

Models of the cooling of Jupiter over

TRANSFER OF MOMENTUM from eddy currents to sustained eastward and westward winds was documented for the earth by Victor P. Starr and his colleagues at the Massachusetts Institute of Technology; the Voyager images suggest that similar transfers feed momentum into the east-west winds on Jupiter and Saturn. In the illustration eddies on a model planet are shown to feed eastward momentum toward the north and westward momentum toward the south. This augments the velocity difference between the planet's mean east-west currents.

JUPITER

SATURN

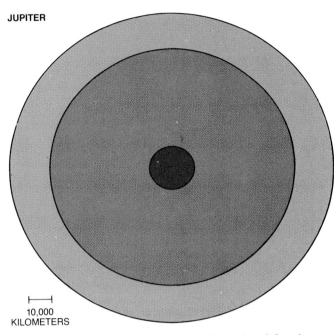

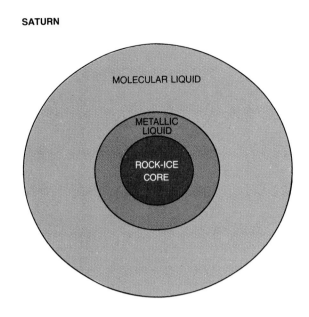

MOLECULAR LIQUID

METALLIC LIQUID

ROCK-ICE CORE

10,000 KILOMETERS

INTERIOR of Jupiter (*left*) and Saturn (*right*) consists of three layers. The outermost layer is a liquid mixture of hydrogen and helium; the hydrogen is molecular. The innermost layer is a core of rock and ice. In the middle layer a pressure in excess of three million earth atmospheres transforms the hydrogen into a liquid mixture of protons and wandering electrons. Thus the hydrogen in the middle layer is a metal. The rotation of each planet makes it somewhat flattened toward the poles. The degree of flattening is greater on Saturn.

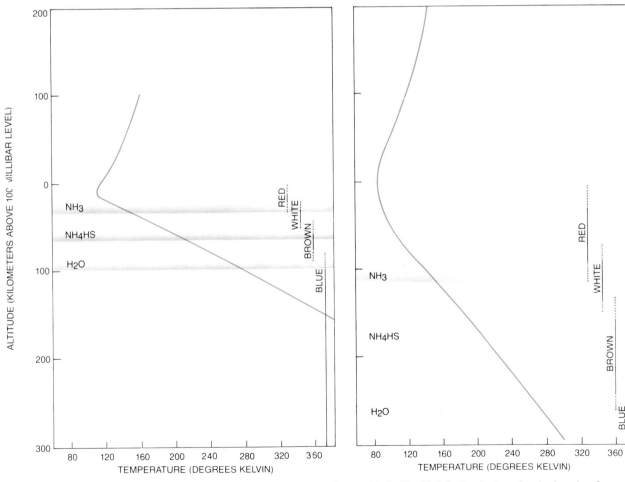

ATMOSPHERE of Jupiter (*left*) and Saturn (*right*) is assumed to have proportions of the various chemical elements much like the proportions in the sun; on that basis each atmosphere can be assumed to include stratified clouds of ammonia (NH_3), ammonium hydrosulfide (NH_4HS) and water. These compounds all form white particles; hence the colors in each atmosphere must have other causes, which have yet to be identified. In the charts each color is assigned a range of altitudes in accord with measurements of cloud-top temperatures. Red clouds, for example, are the coldest and therefore the highest. The cloud tops on each planet lie no higher than the level at which the temperature (*colored curves*) begins to increase with altitude, so that convection currents no longer carry solid (or liquid) particles upward.

the history of the solar system account for all the power that Jupiter is radiating today in excess of the amount it absorbs. Similar models fail, however, to account for about a third of the excess that Saturn radiates. Apparently Saturn has a source of internal heat not included in the calculations, a source Jupiter lacks. In the mid-1970's Salpeter and Stevenson proposed an explanation that is now supported by the Voyager data. A planet incorporating a mixture of hydrogen and helium has two types of energy: thermal and gravitational. If the mixing ratio (that is, the ratio of hydrogen to helium) is constant throughout the mixture, the two types of energy are released together in constant proportions. As the planet cools it contracts. If, however, the mixing ratio changes (for example, if the helium falls through the hydrogen), an additional quantity of gravitational energy is released. Such a process is likely on Saturn because the planet has already cooled to the extent that helium is precipitating at the top of the metallic hydrogen zone.

According to Salpeter and Stevenson, the process is much like ordinary rainfall. When a parcel of the earth's atmosphere is cooled below its saturation point, water condenses and raindrops

fall. The condensation releases the heat that had been added to the water to make it a vapor, and the rate of atmospheric cooling slows. On Saturn the raindrops are helium, and energy is released as the raindrops rub against the hydrogen fluid through which they fall. The process began on Saturn some two billion years ago. Jupiter, being more massive, has not yet cooled to a point where the planet at any level is saturated with helium. Perhaps it is just now reaching that point.

Evidence in support of the hypothesis that helium rain releases energy on Saturn has come from measurements of the atmospheric helium abundance. It is not an easy measurement to make. Helium does not absorb radiation in the infrared, and so the spectrum of the radiation emitted by Jupiter and Saturn does not reveal helium directly. The presence of helium does, however, affect the absorption of infrared radiation by hydrogen. By this means a group led by Rudolph A. Hanel of the Goddard Space Flight Center of NASA and Daniel Gautier of the University of Paris has determined the relative numbers of helium atoms and hydrogen molecules from measurements made by the Voyager spacecraft.

Above the zone of metallic hydrogen

Jupiter turns out to be 10 percent helium, a value not significantly different from the 11 percent of helium in the sun. Saturn turns out to be less than 6 percent helium. If the missing helium has rained downward from the molecular zone to deep in the metallic zone, the heat that would have been released could indeed have maintained Saturn's internal heat source at its present value for the past two billion years. Thus the various inferences about the bulk composition, the internal structure and the history of the cooling of the giant planets all seem to hold together.

Atmospheric Chemistry

As early as the 1930's investigators were identifying lines in the spectrum of sunlight reflected from Jupiter due to the absorption of light by gaseous methane and ammonia. The hydrogen molecule has only weak absorption lines; hence its presence was confirmed only some 30 years later even though its abundance was then inferred to be 1,000 times greater than that of methane and ammonia. The calculated abundances of hydrogen, carbon and nitrogen in the atmosphere of Jupiter and Saturn support the hypothesis that the giant planets

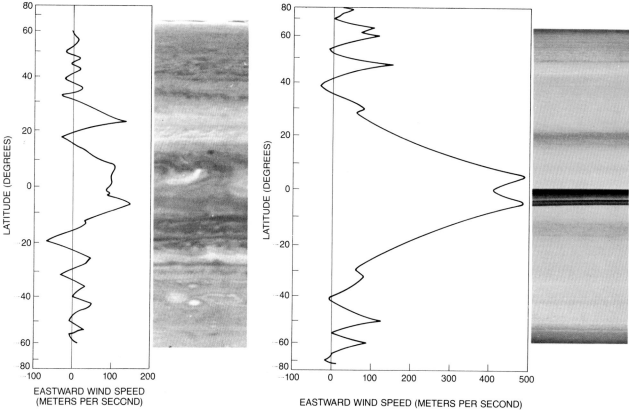

MEAN WIND SPEEDS on Jupiter (*left*) and Saturn (*right*) are compared with images of each planet. Positive numbers are eastward velocities; negative numbers are westward ones. The numbers are wind speeds with respect to the speed of the planet's rotation; they were measured by tracking the motion of cloud features in successive images of each planet made by *Voyager 2*. It emerges from the measurements that each hemisphere of Jupiter has several alternations of eastward and westward currents. Saturn's pattern is simpler but the currents are stronger; in the atmosphere of Saturn 500 meters per second is two-thirds the speed of sound. Eighty years of observations from the earth and recent observations from spacecraft suggest that the cloud colorations at any latitude are changeable but that the winds are much more persistent. The dark band across the equator of Saturn is the ring system of the planet and its shadow on the surface.

are composed of material much like that which condensed to form the sun.

Other gases, however, should also be detected in a mixture of solar composition. Notable among them are water vapor and hydrogen sulfide (H_2S). In the 1970's spectral lines due to the absorption of infrared radiation by water vapor on Jupiter were detected by Harold P. Larson and Uwe Fink of the University of Arizona. Such lines have not been detected on Saturn. Moreover, hydrogen sulfide has escaped detection on both Jupiter and Saturn. Still, both water vapor and hydrogen sulfide gas must have a rather low concentration at the top of the clouds on Jupiter and Saturn, where the temperature is 150 degrees K. (-123 degrees Celsius) or less. Ultimately the vapors may turn out to be present in solar abundances at levels below the clouds.

The infrared spectra of Jupiter's hot spots (holes in the clouds) reveal a rich chemical mixture. In addition to water vapor, phosphine (PH_3), germane (GeH_4), hydrogen cyanide (HCN) and carbon monoxide (CO) have all been detected, and so has heavy methane, that is, methane molecules incorporating the heavy isotope of either hydrogen (hydrogen 2, called deuterium) or carbon (carbon 13). Moreover, ethane (C_2H_6) and acetylene (C_2H_2) have been detected all over the disk of Jupiter and on Saturn as well. Most of these compounds would not be present in an atmosphere rich in hydrogen if the atmosphere were in chemical equilibrium. For example, the various carbon compounds would revert to methane, and the nitrogen would form ammonia.

The obvious source of the chemical disequilibrium is the sun. Specifically, the sun's ultraviolet radiation can break down the dominant chemical species such as methane and liberate free radicals such as CH_2 and CH_3. Lightning in the clouds of Jupiter and Saturn and the impact of electrically charged particles can have the same effect. (The charged particles rain down into the atmosphere from the magnetosphere, the zone above the planet where they are confined by the planet's magnetic field.) The radicals can react with methane molecules to form ethane and acetylene and liberate hydrogen. The composition of the atmosphere of Jupiter therefore reflects a balance between the production and the breakdown of these higher hydrocarbons.

The presence of carbon monoxide in Jupiter's hot spots is more problematic, since water, the potential source of oxygen, is scarce above the clouds, where ultraviolet photons (quanta of ultraviolet radiation) are plentiful. On the other hand, water may be abundant below the clouds, and there it may combine with methane to form carbon monoxide. The conditions of temperature and pressure well below the clouds favor such a reac-

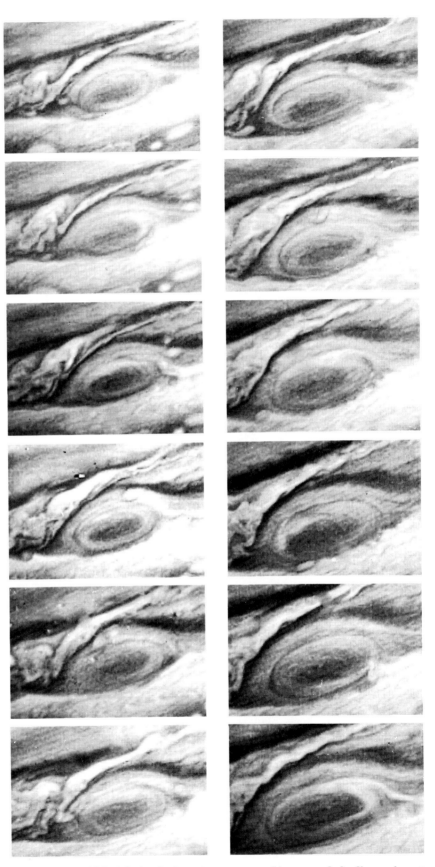

GREAT RED SPOT of Jupiter is shown in a sequence of images made by *Voyager 1* every fourth rotation of the planet, or about once every 40 hours. The spot itself has been observed for three centuries from the earth. It rolls counterclockwise between a westward current to its north and an eastward current to its south. It is some 25,000 kilometers long. In the Voyager sequence structure on a scale of 1,000 kilometers is revealed. Small clouds approach the spot from the east. They circle the spot in six to 10 days. They are partially swallowed. Smaller spots return to the east. In addition to the Great Red Spot, Jupiter has several white ovals.

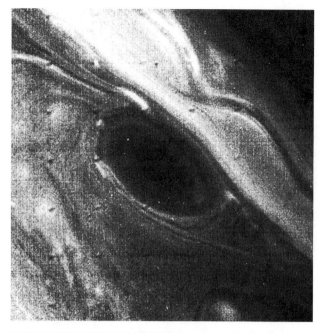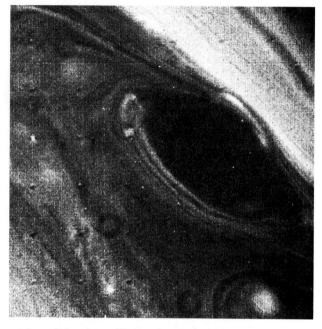

LONG-LIVED SPOT ON SATURN is a brown oval in the north-ern hemisphere that was photographed on August 23 (*left*) and one **rotation of the planet (10 hours) later** (*right*) **by** *Voyager 2.* **It is a fourth the size of the Great Red Spot. Its rotation is clockwise.**

tion. Another possibility is that the oxy-gen above the cloud tops comes not from Jupiter but from sulfur dioxide ejected from the volcanoes of Jupiter's satellite Io, whose activity was revealed in images made by *Voyager 1.*

The solid and liquid particles that constitute the clouds of Jupiter and Sat-urn give further evidence of chemical disequilibrium. The most abundant con-densable vapors in a mixture of solar composition are water, ammonia and hydrogen sulfide. At chemical equilib-rium they form crystals of water ice, of ammonia and of ammonium hydro-sulfide (NH_4HS). Liquid drops of wa-ter and of ammonia-water solutions are also conceivable, as John S. Lewis and his colleagues at the Massachusetts In-stitute of Technology have noted. The problem is that all these condensates are white, whereas the clouds of Jupiter and Saturn are colored.

Ronald G. Prinn of M.I.T. has sug-gested some possible coloring agents. For one thing, molecular sulfur (S_n, where n can take several values greater than 1) forms brown and yellow parti-cles. It therefore becomes all the more vexing that hydrogen sulfide, the parent molecule of the elemental sulfur, has not yet been detected. The challenge for theorists championing sulfur as a color-ing agent is to hide the hydrogen sulfide in the clouds from spectroscopic view but still expose enough of it to solar pho-tons so that S_n can form. Prinn notes that elemental phosphorus has a red col-or resembling that of the Great Red Spot, and it could be formed when ultra-violet photons strike molecules of phos-phine gas. Given the affinity of carbon

atoms for one another, complex organic compounds are also a possible source of color. Unfortunately the spectroscopic identification of compounds in the solid state is difficult because the pattern of vibrations (and the associated absorp-tion of certain wavelengths of light) by the molecules in a solid is blurred by the collisions between neighboring mole-cules. Hence the sources of color on Ju-piter and Saturn remain uncertain.

It can be said, however, that clouds of differing color on Jupiter and Saturn are associated with different levels in the atmosphere. A comparison of images made in visible light and in infrared ra-diation reveals this correlation. The visi-ble images show the colors; the infrared images distinguish cool (and therefore high) clouds from warm (and therefore low) ones. On Jupiter the highest cloud tops are red; the next-highest are white, the ones lower than that are brown and the lowest ones (or perhaps the at-mosphere below the clouds) are blue. Presumably the various compounds responsible for these colors form at dif-ferent levels in response to different temperatures and amounts of sunlight. The white particles that form the major constituents of the clouds should also be layered. Ammonia should be upper-most, then ammonium hydrosulfide and finally water.

Atmospheric Circulation

In the atmosphere of the earth hori-zontal gradients of temperature are the reservoir from which the winds get their energy. The gradients arise because the sun heats the Tropics more than the

poles; then the warm tropical air slides over the colder polar air. This converts gravitational potential energy into ki-netic energy and also transports heat up-ward and poleward.

On Jupiter and Saturn horizontal tem-perature gradients may be less impor-tant than they are on the earth. First of all, the gradient of temperature from the equator to the pole is small on Jupiter and Saturn, at least at the cloud-top levels where the Pioneer and Voyager spacecraft could measure it. On Jupiter, for example, the difference in tempera-ture between the equator and the pole is less than three degrees C. On the earth the difference is 10 times greater, and it is compressed into a distance from the Equator to the pole that is 10 times smaller.

In the second place the atmosphere of Jupiter and Saturn gets half or more of its heat from the interior of the planet. In that regard Jupiter and Saturn resem-ble stars more than they resemble the earth. According to theories of convec-tion employed in stellar models, an in-ternal heat source should produce an adiabatic temperature gradient that is the same along any radius from the cen-ter of the planet to a point on its surface. Moreover, since the part of the atmo-sphere of Jupiter or Saturn receiving sunlight has only a millionth the mass (and therefore a millionth the heat-car-rying capacity) of the planet's interior, one can expect the atmosphere to be in essence short-circuited by the interior. It is as if all points on the surface of the planet were wired to the center by strong conductors, and weak conductors con-nected points on the surface to each other. On this reasoning horizontal temperature gradients should be small.

Still, more sunlight is deposited at the surface of Jupiter and Saturn than at the poles, and measurements made by the Pioneer and Voyager spacecraft show that the rate of infrared emission is roughly independent of latitude and is greater than the rate of absorption of sunlight at all latitudes. Heat is therefore transported poleward. The argument that the interior of each planet is in effect a strong conductor implies that the entire fluid interior of each planet is involved in the poleward transport. A computer model that I and Carolyn Porco have devised at Cal Tech suggests how the transport is maintained. In response to the uneven distribution of sunlight at the surface of the planet small differences in temperature develop in the interior. The differences modulate the rate at which internal heat arrives from below. According to the model, the poleward decrease in the solar heating of the thin sunlit layer is offset by a poleward increase in the upwelling of internal heat. Poleward heat transport in the thin sunlit layer is negligible.

A model based on entirely different assumptions has been published by Gareth P. Williams of the Geophysical Fluid Dynamics Laboratory of the National Oceanic and Atmospheric Administration at Princeton University. Williams began with a mathematical formalism employed in predicting the weather on the earth. The mathematics describes the motions that arise in a fluid atmosphere much thinner than the planet's radius in accord with physical laws such as the conservation of momentum and energy. Williams scaled the radius and the rotation rate of the model planet so that they would characterize Jupiter or Saturn, and he reduced what are called the dissipation parameters, which represent the rate at which an atmosphere loses heat and momentum because of forces such as friction. The reduction of the parameters can be justified on the ground that Jupiter and Saturn have no solid surfaces that winds can rub against. Moreover, a deep, cold atmosphere loses heat extremely slowly, as Peter J. Gierasch, now at Cornell, and Richard Goody of Harvard University pointed out 15 years ago.

Williams also assumed that the atmosphere below the level to which sunlight penetrates on Jupiter or Saturn has negligible effects on the circulation above. In other words, the upwelling of heat from the interior was neglected or assumed to be independent of latitude. In addition the lower boundary of the sunlit layer was assumed to act like a solid surface in that it is undeformable and impermeable. This assumption would be justified if the density of the sunlit layer were substantially less than that of the layer below it. Such a decrease is

found between the warm water at the surface of the earth's oceans and the cooler water below it. The decrease is less likely to be found, however, in a fluid that is convecting heat upward from below.

In short, Williams' model of the atmosphere of Jupiter or Saturn resembles in many ways a model of the atmosphere of the earth. Yet his model proves able to generate the most notable feature of the meteorology of Jupiter or Saturn:

an alternating pattern of east and west winds. Indeed, the model entails a prediction of how the winds are maintained. In the atmosphere of the earth the sustained low-latitude winds blow to the west and the sustained middle-latitude winds blow to the east. Both are maintained by large-scale eddies that transport eastward momentum away from the equatorial latitudes. The eddies include the cyclones and anticyclones that give rise to much of the

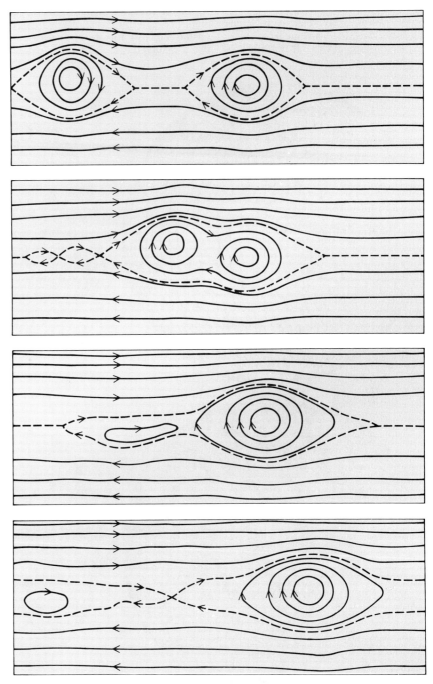

MERGING OF SPOTS on Jupiter and Saturn was simulated with a computer by the author and Pham Giem Cuong at the California Institute of Technology. They propose that each spot is a more or less permanent vortex above an underlying pattern of east-west flow. On their hypothesis the small transient spots on Jupiter and Saturn are buoyant; thus the small spots store gravitational potential energy. The large spots are maintained by swallowing the small ones. The broken lines mark boundaries between each vortex and the atmosphere's laminar flow.

earth's weather. On Jupiter and Saturn the bands of alternating winds are more numerous. Nevertheless, in Williams' model it is eddies that put energy into the east-west winds. Furthermore, the energy driving the eddies on Jupiter, Saturn and the earth comes ultimately (Williams proposes) from the same source: the temperature gradient from the equator to the poles that is maintained by solar heating.

Voyager Measurements

The best measurements of the winds on Jupiter and Saturn come from tracking the position of cloud features in suc-cessive images made by the Voyager spacecraft. The investigators responsible for making these measurements included Reta Beebe, Garry E. Hunt, Jim L. Mitchell and me. For Jupiter our first goal was to define the mean wind velocity at each degree of latitude by averaging the velocities of all the features we could identify at that latitude. At most latitudes 50 to 100 high-contrast features could be tracked. Saturn has fewer such features and so fewer measurements could be made.

One remarkable result was the agreement between the measurements based on the Voyager images of Jupiter and the measurements based on some 80 years of observations from the earth. The changes over the 80 years seem to have been mostly in the coloration of the clouds at given latitudes. At times the east-west winds were unobservable from the earth, but apparently they did not change.

A second result has to do with how the winds deviate from their mean velocity at a given latitude. On Jupiter the mean eastward velocities are no greater than 130 meters per second. The deviations from this mean, which correspond to eddies, are typically 20 meters per second. The deviations have a systematic tilt. For example, at places where the mean eastward velocity increases with lati-

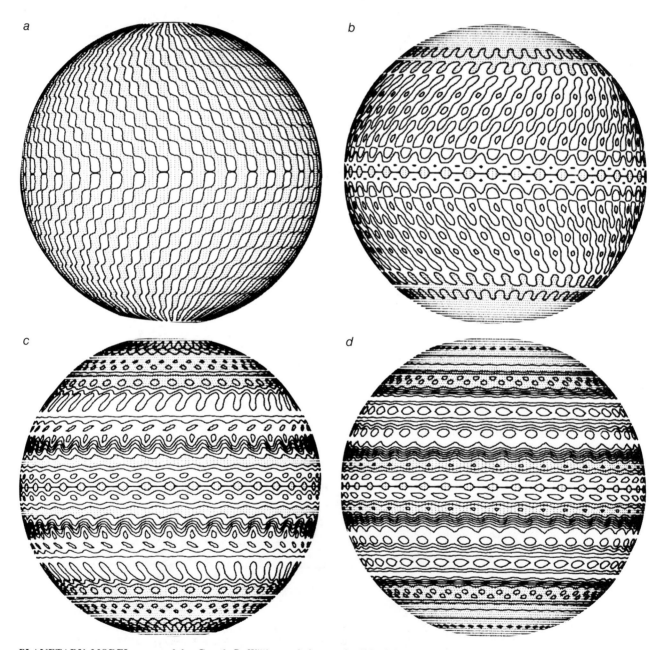

PLANETARY MODEL proposed by Gareth P. Williams of the Geophysical Fluid Dynamics Laboratory of the National Oceanic and Atmospheric Administration assumes that the processes shaping the weather on the earth also shape the weather on Jupiter and Saturn. Specifically the uneven distribution of solar radiation across the disk of the planet entails eddy currents in a thin sunlit layer of the atmosphere. The momentum of the eddies then fuels eastward and westward currents. Trains of spots arise in the model but isolated ovals do not. The illustration shows the state of Williams' computer simulation after 4.6 days (*a*), 23 days (*b*), 46 days (*c*) and 73.3 days (*d*).

tude, the vectors that represent the velocity (and therefore the momentum) of the deviations tend to be tilted toward the northeast. Hence they feed velocity and momentum northward into latitudes where the mean velocity and momentum are already greater. This is precisely the mechanism by which eddies sustain the mean winds on the earth, and it is also the mechanism that underlies Williams' model. Jupiter's seemingly chaotic eddies have more order than one might suspect.

There are nonetheless some important quantitative differences between Jupiter and the earth. On the one hand, the mean velocities of the winds and the velocities of the eddies are greater on Jupiter. Indeed, the rate at which kinetic energy is transferred from eddies to sustained east-west winds is 10 times greater on Jupiter than it is on the earth per unit area on the surface of the planet. On the other hand, the rate at which thermal energy is made available to the atmosphere of Jupiter for possible conversion into kinetic energy of winds is 20 times less than it is on the earth, again per unit area, because Jupiter is cooler. Thus the efficiency with which the atmosphere converts thermal energy into kinetic energy seems to be much greater on Jupiter than on the earth.

And yet one must be careful in making any general conclusion about the efficiency of Jupiter or Saturn's energy cycle because these cycles involve transformations of energy that have not yet been measured. For example, a mass of cold air above a mass of warm air represents a certain amount of gravitational potential energy, which is released when the cold air sinks. Such releases occur inside the clouds of Jupiter and Saturn, and thus they are hidden from view. On the earth the net efficiency of the atmosphere's energy cycle is about 1 percent. That is, about 1 percent of the solar power absorbed by the earth fuels the large-scale motions of the atmosphere before friction in the atmosphere converts it back into heat that the earth radiates into space. The rest is radiated into space without ever being converted into kinetic energy. The net efficiency of Jupiter and Saturn is unknown.

Coaxial Cylinders

I myself suspect that the great depth of the fluid interior cannot be neglected in modeling the atmospheric dynamics of Jupiter or Saturn. Laboratory experiments, including those done by Geoffrey I. Taylor at the University of Cambridge in the 1920's and those being done by F. H. Busse at the University of California at Los Angeles today, show that small-scale turbulent motions (that is, eddies) in a rapidly rotating fluid align themselves in columns parallel to the axis of rotation. At any level in each column the motion in a plane perpendic-

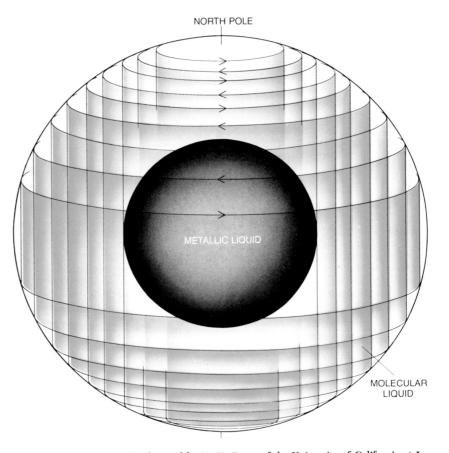

NORTH POLE

METALLIC LIQUID

MOLECULAR LIQUID

ALTERNATIVE MODEL advanced by F. H. Busse of the University of California at Los Angeles and supported by the author proposes that the east-west winds in the cloud decks of Jupiter and Saturn are the visible sign of a pattern of rotation extending through the fluid interior of each planet. The model is based on the experiments of Geoffrey I. Taylor at the University of Cambridge and those of Busse, which suggest that in a rotating fluid planet that has been well mixed by convection the sustained motions are those of nested fluid cylinders. A discontinuity in the density of the planet's interior at the top of the zone of metallic hydrogen interrupts the innermost cylinders. The small asymmetries between the northern and southern hemispheres in the wind profiles shown on page 64 are not inconsistent with the model.

ular to the axis is the same. The columns span their container and resist stretching or compression along the axis. Therefore if the container is spherical, each column remains at a fixed distance from the axis of rotation. Busse's experiments and calculations suggest that a sustained pattern of flow develops in the fluid as a result of the columns. In a liquid of constant density and low viscosity the pattern turns out to be one in which coaxial cylinders of fluid rotate at different velocities about their common axis, which is the axis of the fluid's mean rotation. In a liquid of greater viscosity or stratification into layers of varying density other patterns are possible.

The relevance of these laboratory experiments to Jupiter and Saturn depends, then, on the distribution of density in each of the planets. A marked increase in density below the sunlit layer of the atmosphere, as postulated by Williams, implies that each planet is the large-scale analogue of a stratified laboratory fluid. Columns of the type found by Taylor and Busse do not cross the interface between two layers of differing density, and so the sunlit layer can be

decoupled from the interior. On the other hand, the convection of heat in the interior of Jupiter and Saturn should lead to an adiabatic gradient of temperature and a well-mixed interior. This implies that each planet is the analogue of a laboratory fluid of constant density.

According to most models of Jupiter and Saturn's interior, the adiabatic zone ends in the clouds. Hence the steady eastward and westward winds we have measured in the clouds of Jupiter and Saturn may turn out to be the visible sign of the motion of coaxial cylinders that extend all the way through each planet's fluid interior. To be sure, the winds might get their energy from eddies, as is suggested by Williams' model and supported by data from the Voyager spacecraft. If the winds do represent the motion of cylinders, however, the inertia that supports the winds would be immense. Once the coaxial cylinders are set in motion they might well remain in undisturbed rotation throughout the 80 years of observations made first by earth-based astronomers and then by the Voyager instruments.

The coaxial cylinders do imply that

the profiles of wind velocity v. latitude in the northern and southern hemispheres of Jupiter and Saturn should be symmetrical about the equator. The symmetries need not, however, extend to high latitudes, because the jump in density at the interface between molecular and metallic hydrogen inside each planet decouples the inner coaxial cylinders. For Jupiter the decoupling should affect latitudes greater than 40 to 45 degrees North and South. For Saturn it should affect latitudes greater than about 65 degrees. In profiles produced from Voyager data the requisite symmetries are present: the northern and southern hemispheres of Jupiter and Saturn each show about three complete cycles of alternating eastward and westward winds between the equator and the latitudes affected by decoupled cylinders. Small departures from symmetry, such as the ones at 25 degrees north and south on Jupiter, where eastward jets are found, may be due to a nonadiabatic temperature gradient in the cloud layers. Larger departures would imply a nonadiabatic interior and would constitute a disproof of the hypothesis.

It is difficult at present to choose between a model of Jupiter and Saturn in which the great depth of the fluid interior is immaterial and one in which the depths are crucial. After all, even the data recorded by the Voyager spacecraft pertain only to the cloud tops. The spacecraft *Galileo,* which is scheduled for launching in the late 1980's, will probe the atmosphere of Jupiter more deeply. It is hoped the probe will measure winds at the level of the cloud bases and will determine if the adiabatic zone extends that high. Meanwhile some indirect strategies may prove useful.

One such strategy exploits the differences between Jupiter and Saturn. In particular, the winds at the cloud tops on Saturn are three to four times stronger than the winds on Jupiter. (The eastward wind speeds at the equator of Saturn are almost 500 meters per second, or more than 1,000 miles per hour.) The eastward and westward currents are broader than on Jupiter, and fewer large oval structures are found. The possible causes of these differences cannot be many. Perhaps, for example, it is significant that the zone of metallic hydrogen lies much deeper in Saturn than in Jupiter. The corresponding increase in the depth of the molecular hydrogen zone might give coaxial cylinders that rotate in opposite directions a greater spacing and greater relative velocities. On the surface of Saturn (as compared with the surface of Jupiter) gravity is weaker, the flux of heat is lower and seasonal changes are greater. (The last follows because Saturn's axis of rotation is more inclined than Jupiter's with respect to the plane of its orbit around the sun and because the rings of Saturn cast a shadow on the surface that changes its position seasonally.) As a result of the weaker gravity Saturn's clouds are thicker and as a result of the lower flux of heat the small-scale convective motions of the atmosphere might be weaker. The implications of the differences at the surface of each planet for large-scale motions in the atmosphere are, however, unclear.

Long-lived Ovals

Another strategy is to examine Jupiter and Saturn's atmospheric flow patterns other than the eastward and westward winds and see what assumptions about the deep interior are compatible with the observations. In particular, the Great Red Spot of Jupiter and other long-lasting ovals on Jupiter and Saturn are a unique and possibly diagnostic feature of the giant planets. The ovals themselves are relatively slow-moving. For example, the Great Red Spot moves westward only a few meters per second, although the winds around it reach speeds of 100 meters per second. In fact, all the ovals on Jupiter and Saturn seem to rotate like ball bearings between adjacent eastward and westward currents. Each rotation takes only a few days.

The ovals are also enduring: they can last for decades and even centuries. The eddies in the oceans and the atmosphere of the earth are less enduring by orders of magnitude. For example, the eddies in the Atlantic tend to drift westward until they merge with the Gulf Stream off the east coast of North America. Their lifetimes are measured in months and sometimes years. The eddies in the atmosphere of the earth are of several types. The most enduring ones seem to be trapped in place near features of the surface such as mountain chains or boundaries between a continent and an ocean. On Jupiter and Saturn there is no such topography.

At least two proposals have been put forward that account for many of the properties of the ovals. Each proposal seeks only to show how an isolated vortex can endure in the midst of a pattern of alternating eastward and westward currents. It seeks, in other words, to show that the configuration is stable even if small perturbations of the configuration occur. The proposals do not account for how the vortexes arise.

Tony Maxworthy and Larry G. Redekopp of the University of Southern California have proposed that a long-lived oval represents a "solitary wave," that is, a self-sustaining wave with a single crest instead of a train of crests and troughs. Such a wave is a fluid-dynamical curiosity dating back to the 19th century; in Maxworthy and Redekopp's hypothesis it is a single north-south displacement of flow lines that otherwise lie east-west. The models Maxworthy and Redekopp have published often bear a striking resemblance to the Great Red Spot. Moreover, those models imply that a planetary solitary wave can exist only in a pattern of east-west flow that is unstable to certain perturbations. This may explain why isolated ovals do not spontaneously appear in Williams' models, in which the east-west flows are not unstable. In Williams' models the closest analogues to isolated ovals are trains of spots strung out at constant latitude. Such patterns do appear on Jupiter, but they are distinct from the large, isolated ovals. When two solitary waves meet, however, they simply pass through each other. When two ovals meet on Jupiter or Saturn, they sometimes merge.

The second proposal, made by me and Pham Giem Cuong of Cal Tech, depends on assuming that the east-west pattern of flow in the clouds is part of a much deeper pattern, perhaps one of rotating coaxial cylinders. On that hypothesis it can be shown that stable vortexes can exist in an east-west flow that is also stable. According to this proposal, such a vortex extends downward only to the top of the adiabatic zone, wherever that may lie. We tested the stability of the ovals in our computer model by introducing large and small perturbations, by forcing two ovals to collide and by feeding small ovals to the larger ones. Given the right east-west flow under them, the ovals are quite robust: they survive rather large perturbations. Moreover, the large spots can grow by consuming the smaller spots. It is likely that on Jupiter and Saturn the smaller, transient spots derive energy from their buoyancy.

So far Williams' model of Jupiter and Saturn, which derives from models of the earth, is the only proposal that is complete, in the sense that it has sources and sinks of energy and the eastward and westward currents arise spontaneously. To varying degrees the other models take much of the pattern as presupposed. For example, the models of the long-lived ovals presuppose the basic east-west flows.

The fundamental issues remain unresolved. How deep do the visible patterns of flow on Jupiter and Saturn extend? How important is solar heating of the atmosphere, as opposed to internal heating? How is the density of the atmosphere stratified below the cloud tops? The issues are particularly challenging because an all-encompassing computer model of Jupiter or Saturn is impractical. One simply cannot incorporate into the same mathematical description of a planet the small, transient atmospheric eddies that are crucial in Williams' hypothesis and the slow, large-scale internal responses to uneven solar heating that occur in the hypothesis advanced by Porco and me. The scales of size and time for the two phenomena are just too different. Computer models will be important, but as we come to understand the giant planets clever thinking and insight will be needed even more.

7

The Galilean Moons of Jupiter

by Laurence A. Soderblom
January, 1980

In viewing the four largest Jovian satellites last year the Voyager spacecraft increased to nine the number of earthlike bodies that can be closely compared to gain an understanding of how they evolved

One of the most spectacular scientific adventures of all time began on March 5 of last year. In a period of some 30 hours the spacecraft *Voyager 1* flew past the giant planet Jupiter and returned closeup pictures of three of the planet's four largest moons: Io, Ganymede and Callisto. First observed by Galileo in 1610, the four moons are commonly called Galilean. Detailed pictures of the fourth Galilean satellite, Europa, were made later (on July 9) by *Voyager 2,* which also explored the hemispheres of Ganymede and Callisto that had not been visible to its sister craft. Perhaps the most dramatic discovery made by *Voyager 1* was that on Io volcanic eruptions were in progress. *Voyager 2* therefore devoted nearly 10 hours to observing Io in a "volcano watch" inspired by the earlier pictures.

Both spacecraft also had distant views of Amalthea, an asteroidlike body discovered only 88 years ago and until recently thought to be Jupiter's innermost satellite. Voyager images have now disclosed, however, that the planet has another tiny satellite near the outer edge of a faint ring resembling one of Saturn's rings. This 14th satellite of Jupiter, which is only a few dozen kilometers in diameter and is about halfway between the surface of the planet and the orbit of Amalthea, was discovered by David Jewitt and G. Edward Danielson of the California Institute of Technology. It has temporarily been designated 1979J1.

Amalthea and the four Galilean moons (and probably 1979J1 as well) travel in circular orbits lying in Jupiter's equatorial plane and therefore constitute the "regular" satellite system of the planet. The remaining eight confirmed Jovian moons are much smaller and move in irregular orbits scattered well beyond those of the inner six. From close examination of the several hundred pictures of the Galilean satellites and Amalthea returned by the two Voyagers one can infer a great deal about their histories, relative ages and the nature of the geological processes by which they have evolved.

The use of the term satellites to describe the Galilean moons understates their importance to students of planetary science. Io, Europa, Ganymede and Callisto belong to the family of objects designated terrestrial, a family that includes Mercury, Venus, the earth, the earth's moon and Mars. The Galilean satellites are similar to the bodies of the inner solar system in both size and composition, so that both groups of objects should have evolved by comparable processes and on similar time scales. As a result of the Voyager missions the number of earthlike objects with which one can test theoretical models of planetary evolution has doubled. One can now appreciate that the planets of the inner solar system occupy only a small part of the spectrum of characteristics and evolutionary possibilities open to such objects.

It was known from studies done with telescopes on the earth that the Galilean satellites exhibit certain regular trends, such as decreasing density and increasing size with distance from Jupiter, trends remarkably like those in the solar system as a whole. Of Amalthea little can be told from the earth except that it is no more than a few hundred kilometers in diameter, dark and red. Io, the next satellite out from Jupiter, was known to have about the same size and density (3.5 grams per cubic centimeter) as the earth's moon. Unlike the other three Galilean satellites, Io shows no trace of water in its infrared reflection spectrum. Because of its brilliant orange red appearance and the steep drop in its reflection spectrum toward the ultravio-

let it was suspected of having a surface rich in sulfur.

Europa, second out from Jupiter of the large Galilean satellites, is also about the size of the earth's moon but is far brighter: it reflects nearly 70 percent of the sunlight striking it compared with 7 percent for the moon (and 35 percent for the earth). Spectra made at near-infrared wavelengths indicated large amounts of water ice on the satellite's surface. Estimates of Europa's density (about three grams per cubic centimeter) suggested that it might have a shell of ice and liquid water as much as 100 kilometers thick.

The third and fourth members of the Galilean satellite system, Ganymede and Callisto, are rather similar. Both were known to be about the diameter of the planet Mercury and to have a density of about two grams per cubic centimeter. (Mercury and the earth are about 5.5 grams per cubic centimeter.) One could therefore speculate that Ganymede and Callisto are even richer in water than Europa is, assuming that all three consist chiefly of water and typical silicates (silicon oxides). Ganymede, however, reflects 50 percent of the sun's light compared with only 20 percent for Callisto and shows more evidence of water or ice on its surface.

The Voyager Trajectories

As Amalthea and the four Galilean satellites move in their orbits they keep one face constantly toward Jupiter, just as the earth's moon keeps one face toward the earth. Their orbital periods

CALLISTO, the outermost Galilean moon, was photographed on March 5, 1979, at a distance of between 337,000 and 364,000 kilometers by the television cameras aboard the spacecraft *Voyager 1.* The other three Galilean moons, in the order of their distance from Jupiter, are Io, Europa and Ganymede. Jupiter has 14 moons in all, including a tiny new one discovered by *Voyager 2* and temporarily designated 1979J1. The new moon circles Jupiter within the orbit of another small moon, Amalthea, which in turn circles the planet inside the orbits of the Galilean satellites. The orbits of the remaining eight Jovian moons lie outside those of the Galilean satellites. Callisto is about the size of Mercury. It circles Jupiter once every 16.69 days at a distance of 1.8 million kilometers. In this mosaic the resolution, equivalent to the width of a pair of television scan lines, is seven kilometers. The circular feature near the left limb, or edge, is about 600 kilometers across. Concentric rings extend outward for another 1,500 kilometers.

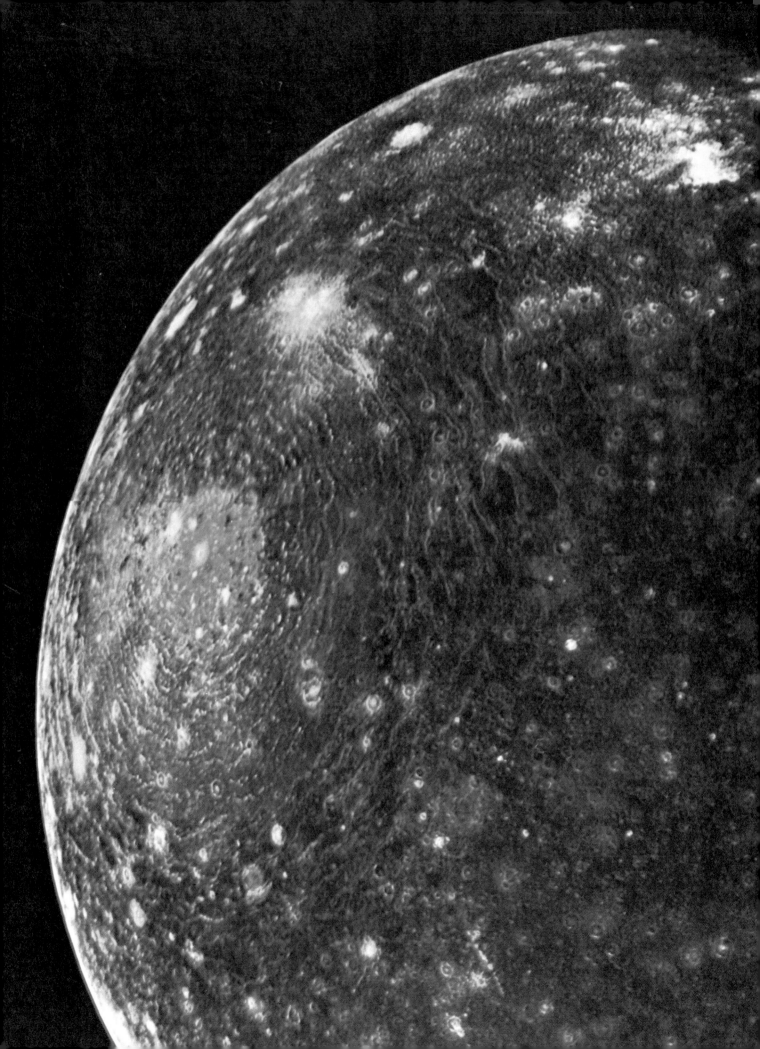

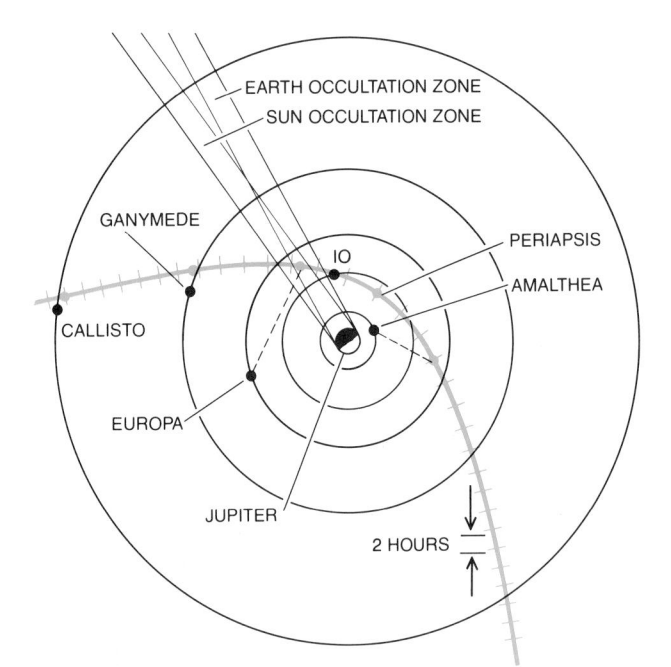

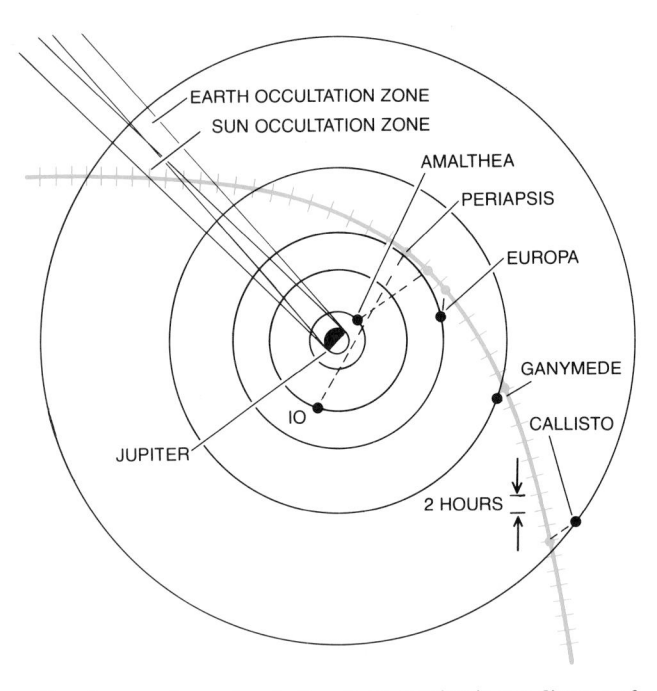

TRAJECTORIES OF THE TWO VOYAGERS were selected to optimize their coverage of the Galilean satellites. *Voyager 1* (*left*) went within 277,000 kilometers of Jupiter on March 5. *Voyager 2* (*right*) did not go nearly as close to the planet, passing it at a distance of 650,000 kilometers on July 9. In these diagrams each satellite is at the position it occupied when spacecraft made its closest approach.

range from 12 hours for Amalthea to 16.69 days for Callisto. The Voyager trajectories took advantage of this synchronous rotation of the Galilean satellites and their rapidly changing positions to optimize pictorial coverage of their surfaces. As *Voyager 1* approached Io the spacecraft took pictures of the satellite's outward-facing and trailing hemispheres, flew under its south pole and proceeded to close encounters with Ganymede and Callisto, viewing their Jupiter-facing hemispheres and flying past them at high latitudes so that their north poles could be examined.

Voyager 2 encountered Callisto and Ganymede before it reached Jupiter and hence was able to photograph the opposite, outward-facing hemispheres and to

view the south-polar region of Ganymede. As a result 80 percent of the surface of Ganymede and Callisto were covered with a resolution of about five kilometers or better. (Resolution is defined as the width of two television scan lines.) *Voyager 2* also photographed about a fourth of the surface of Europa with a similar resolution. Since Amalthea and Io revolve around Jupiter rapidly compared with the time each spacecraft was near the planet, they could be photographed at all longitudes with intermediate resolution (about 20 kilometers per line pair). Io's short orbital period (1.77 days) also made it possible for the two Voyagers to inspect the satellite repeatedly at many longitudes during the few days near each encounter. This provided a rather complete inventory

and characterization of the plumes that marked the volcanic eruptions on the satellite.

The most exciting aspect of the satellites of Jupiter is what can be learned from comparing them. A great variety of geological processes and evolutionary rates are dramatically recorded in their appearance. I shall now describe Amalthea and the four Galilean satellites, in order of their distance from Jupiter, as revealed by the Voyager missions.

Amalthea

Amalthea is in a class by itself. Although it is smaller than the Galilean satellites by a factor of about 10, it is about 10 times larger than Mars's tiny

SATELLITE	DIAMETER (KILOMETERS)	MEAN DISTANCE FROM JUPITER (KILOMETERS)	ORBITAL PERIOD (DAYS)	BULK DENSITY (GRAMS PER CUBIC CENTIMETER)	MASS (MOON = 1)	CLOSEST APPROACH (KILOMETERS)		BEST RESOLUTION (KILOMETERS PER LINE PAIR)	
						VOYAGER 1	*VOYAGER 2*	*VOYAGER 1*	*VOYAGER 2*
AMALTHEA	155 × 270 (±8)	109,900	.49	?	?	420,100	558,270	7.8	11
IO	3,638 (±10)	350,200	1.77	3.53	1.21	18,640	1,127,920	1	21
EUROPA	3,126 (±10)	599,500	3.55	3.03	0.66	732,270	204,030	33	4
GANYMEDE	5,276 (±10)	998,600	7.16	1.93	2.03	112,030	59,530	2	1
CALLISTO	4,848 (±10)	1,808,600	16.69	1.79	1.45	123,950	212,510	2.3	4

PHYSICAL AND ORBITAL CHARACTERISTICS of Amalthea and the Galilean satellites are tabulated with the closest approach of the two Voyagers and the resolution of the best images. Each spacecraft bore two Vidicon cameras with lenses of different focal length: 200 and 1,500 millimeters. Color images were produced by making sequential exposures through orange, green, blue, violet and ultraviolet filters. Each image consists of 800 scanning lines with 800 pixels (image units) on each line. The two Voyagers returned 35,000 pictures in all. All the pictures in this article were prepared at the Jet Propulsion Laboratory of the California Institute of Technology.

RADIATING FLOW PATTERNS ON IO around a dark caldera, the crater of a collapsed volcano, appear in this picture made by *Voyager 1* shortly before its closest approach. The resolution is about two kilometers. At that point the first of the volcanic plumes had not yet been recognized. The calderas gave Voyager investigators their first clue to the absence of the impact craters they had expected to see long before *Voyager 1*'s closest approach. Even in the pictures of the highest resolution, 600 meters, no impact craters were ever seen. Their absence indicates that the volcanism of the satellite is erasing features as large as one kilometer across in less than a million years.

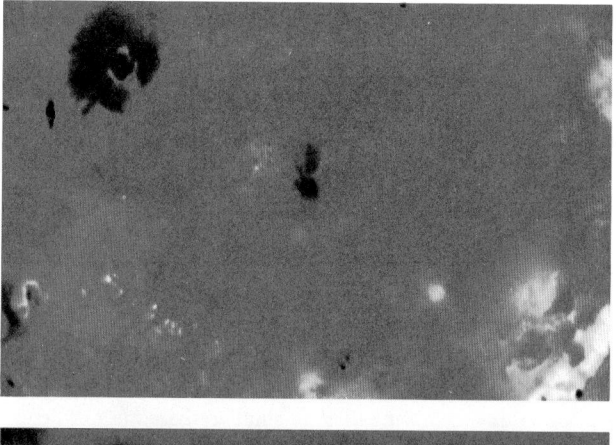

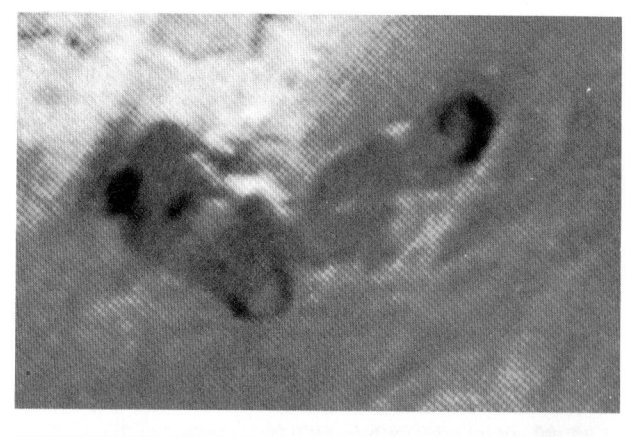

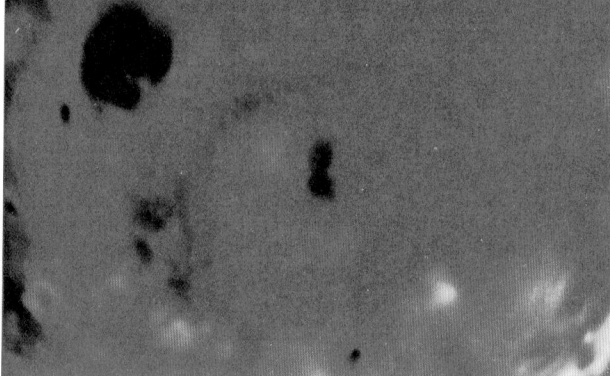

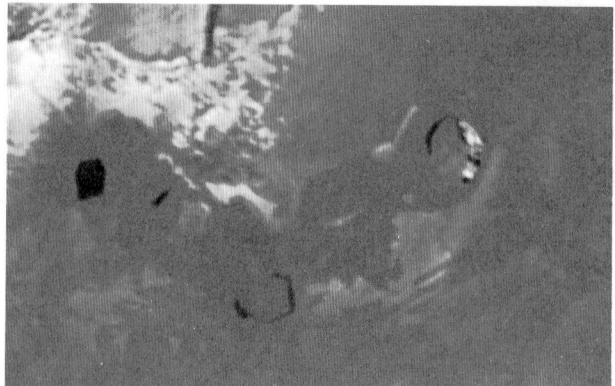

LARGEST ACTIVE VOLCANO ON IO, the source of Plume 1, is shown in the view at the top made by *Voyager 1*. The heart-shaped pattern extends about 1,200 kilometers in its longest dimension. When the same region was photographed by *Voyager 2* (*bottom*), the heart-shaped pattern had been replaced by a more symmetrical one. Other *Voyager 2* images showed that the eruptions had stopped. The *Voyager 2* image, made at a greater distance, has a resolution of 24 kilometers compared with one of seven kilometers for *Voyager 1* image.

THREE CALDERAS ON IO are shown in images made six hours apart as *Voyager 1* closed in on the satellite from a distance of 374,000 kilometers to 130,000 kilometers. In the first view (*top*) the floors of all three calderas show up as black. In the second view (*bottom*) luminous bluish white patches have developed in the caldera that is farthest to the right. A possible explanation is that liquid sulfur dioxide has leaked out of the interior of the satellite and on reaching the surface has exploded into a large cloud of ice crystals and gas.

satellites Phobos and Deimos, which have been intensively examined by spacecraft missions to Mars. Amalthea is therefore the first in a class of intermediate-size objects, comparable to many asteroids, to be explored at close range. Of irregular ellipsoidal shape, its long axis (about 270 kilometers long) is pointed at Jupiter and its short axis (155 kilometers long) is at right angles to the plane of its orbit. Voyager imaging data show that it reflects about 50 percent more red light than violet light, confirming observations made from the earth. Although its overall albedo, or reflectivity, is about 5 percent, a few patches are as much as three times brighter [*see bottom illustration on next page*].

Rudolf A. Hanel of the Goddard Space Flight Center of the National Aeronautics and Space Administration and his colleagues on the Voyager infrared experiment discovered that Amalthea is warmer than it would be if it were simply absorbing and reradiating solar radiation and radiation reflected from Jupiter. The excess heat may be generated by electric currents traveling along the lines of force in Jupiter's magnetic field or by the bombardment of particles trapped in the planet's radiation belt. The irregular shape of the satellite implies a substantial internal rigidity and a low content of volatile substances.

Io

Io, the innermost Galilean satellite, was generally expected to have an ancient cratered surface much like that of the earth's moon. There were, however, a few dissenters from this view. In a striking example of scientific prediction, published just three days before the close approach of *Voyager 1* to Io, Stanton J. Peale of the University of California at Santa Barbara and Patrick M. Cassen and Ray T. Reynolds of the Ames Research Center of NASA pointed out that since the satellite is subjected to resonant gravitational forces exerted by its sister satellites, mainly Europa, its orbit would be distorted out of the circular. Io would move in and out slightly in Jupiter's powerful gravitational field, with the result that it would be repeatedly flexed by tidal forces and an enormous amount of frictional heat would be dumped into its interior. And since the heat would ultimately have to be dissipated through the surface of the satellite, Peale, Cassen and Reynolds speculated, "widespread and recurrent surface volcanism would occur."

The prediction was borne out, but not immediately. The first images of Io with detectable surface markings showed small dark spots, some of which were surrounded by faint rings. The first impression was that these were the expected impact craters. The impression soon had to be revised when pictures of higher resolution showed no impact craters

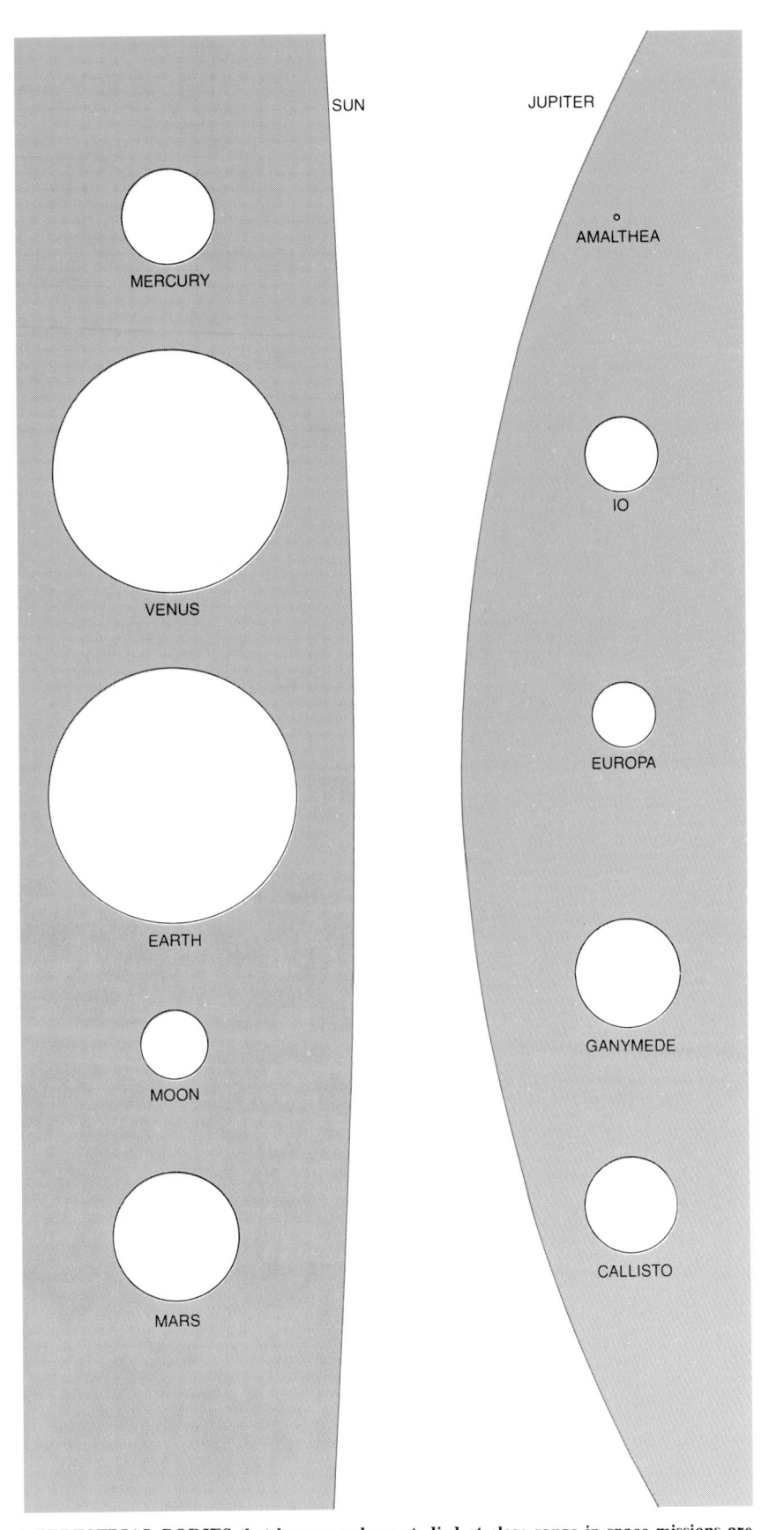

TERRESTRIAL BODIES that have now been studied at close range in space missions are Mercury, Venus, the earth, the earth's moon, Mars and the four Galilean moons of Jupiter. These bodies are designated terrestrial because of their similarity in size and composition to the earth. The five terrestrial bodies of the inner solar system are drawn to the same scale at the left in the order of their distance from the sun. The shallow arc around them represents the limb of the sun. The Galilean moons are drawn to the same scale against the limb of Jupiter and in order of their distance from the planet. At this scale Amalthea is little more than a dot.

JUPITER, IO AND EUROPA were photographed by *Voyager 1* on February 13, 1979, nearly a month before the spacecraft made its closest approach to the planet. The distance was 21.6 million kilometers; the resolution is 390 kilometers per pair of television scan lines. Io appears against the disk of Jupiter, directly above the Great Red Spot. Europa is at the right.

at all! At maximum resolution (600 meters per line pair) a crater as small as one kilometer in diameter should have been visible. Assuming that on Io the rate of formation of impact craters larger than one kilometer is not drastically different from rates typical for the inner solar system, some process on Io's surface must be obliterating the impact craters in a period of something less than a million years.

Io's surface shows a host of bizarre land forms, including sinuous scarps and faults [*see bottom illustration on page 78*]. Pictures returned just before *Voyag-*

er 1's closest approach to Io finally offered a clue to the process that might be renewing the satellite's surface. One such picture shows a circular depression about 50 kilometers in diameter, rimmed by a cliff and surrounded by a radiating pattern of flow [*see top illustration on page 74*]. The feature resembles a volcanic form seen on other terrestrial bodies: a caldera, or large collapse crater, created when lava flows out of or is withdrawn from a magma chamber so that the surface above it collapses. More than 100 such features larger than about 25 kilometers have been identified in the

hemisphere seen in high resolution. The flows radiating from the presumed volcanic centers are multicolored: black, yellow, red, orange and brown. To account for the observations members of the Voyager teams have suggested that the flows consist of everything from sulfur-colored basalt to pure sulfur; some have even suggested the existence of sulfur lakes and oceans.

If the surface of Io is so young, and if volcanic activity was as recent as the pictures indicate, that activity should be continuing even today. Those of us on the Voyager imaging team did not dream, however, of detecting active volcanism on Io. The probability of seeing an active volcano on the earth at a comparable distance with cameras like those on the Voyagers is extremely small.

A few days after the encounter of *Voyager 1* with Jupiter, Linda A. Morabito, an engineer at the Jet Propulsion Laboratory of the California Institute of Technology, was examining some images of Io taken at long range for the purposes of navigation. The pictures had been overexposed to bring out the stars in the field behind the satellite. As Morabito was examining one picture she noted a large bright form shaped like an umbrella off the limb, or edge, of Io's southern hemisphere. When efforts to explain away the form as an artifact failed, the conclusion was reached that it was real. Evidently an enormous cloud was arching 270 kilometers above the surface of Io [*see top illustration on page 78*].

After the reality of the cloud was accepted a search revealed that there were no fewer than eight active volcanoes on Io throwing up plumes from 70 to 300 kilometers high with velocities ranging up to a kilometer per second. Some of the plumes are highly symmetrical. In most instances the ejected material rises to a height of about 100 kilometers and forms the umbrella-shaped cloud as it spreads out and plunges back to the surface. Some of the plumes exhibit a large diffuse globe in the ultraviolet and an inner core in the visible wavelenths [*see illustration at bottom left on page 74*]. The inner core, which is often asymmetrical, may consist of solid particles ejected on ballistic trajectories. The ultraviolet globe is possibly the condensate of a symmetrically expanding gas. Of the eight volcanic plumes observed by *Voyager 1* seven were still there when *Voyager 2* arrived four months later. Evidently the plumes erupt continually and last from a few months to a few years.

Another interesting phenomenon on Io is the presence of white or bluish white bright patches along scarps and faults observed in many areas but notably in the south-polar region. Such features are diffuse, are sometimes variable and apparently obscure the surface below, and so it was first suggested that

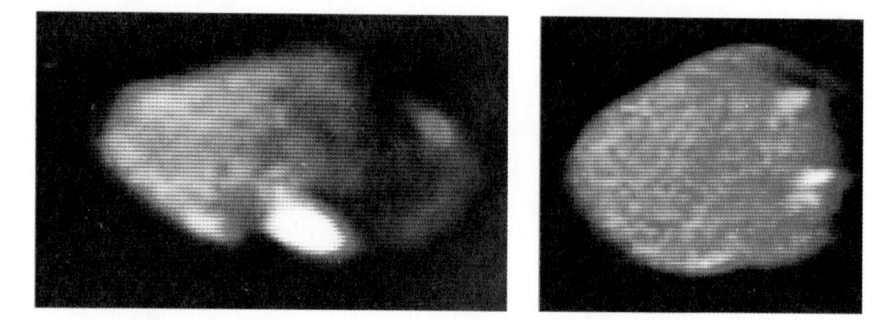

BEST VIEWS OF AMALTHEA were recorded by *Voyager 1*. This asteroidlike body is about 270 kilometers on its long axis and 155 kilometers on its short axis. The picture at the left, made from a range of 695,000 kilometers, has a resolution of 13 kilometers per line pair. Picture at right, made from a distance of 425,000 kilometers, has a resolution of eight kilometers.

they might be clouds (or material deposited from clouds) produced by a gas leaking out of the interior of the satellite and condensing into some kind of snow. One of the key observations that has led to a preliminary explanation of both the plumes and the isolated white patches along scarps is the discovery by John

Pearl of the Goddard Space Flight Center and members of the Voyager infrared experiment that Io has a tenuous atmosphere of sulfur dioxide. From measurements of infrared absorption it has been estimated that during the day this atmosphere has a pressure of about a tenth of a microbar (a ten-millionth of

the atmospheric pressure at sea level on the earth).

Another infrared-absorption feature, discovered in the course of telescopic observations made from the earth by Dale P. Cruikshank of the University of Hawaii at Manoa, was interpreted at about the same time by a number of

 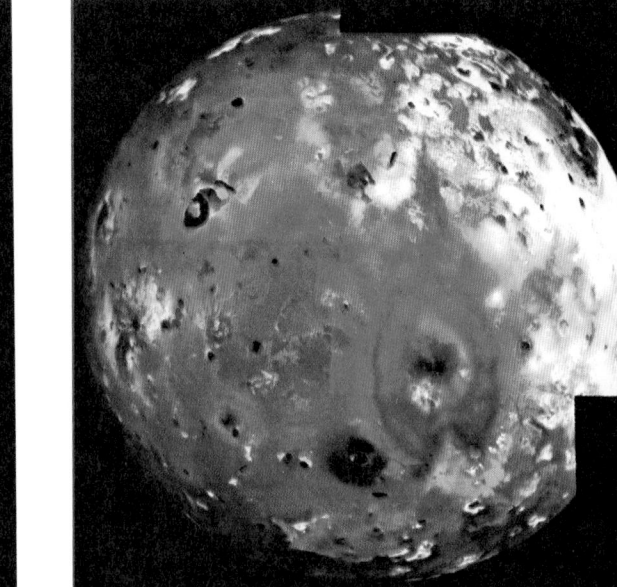

TWO FACES OF IO were recorded about 11 hours apart as *Voyager I* approached and then overtook this innermost Galilean moon. The picture at the left, made from a distance of about 860,000 kilometers, shows the hemisphere of Io that always faces away from Jupiter as the satellite circles the planet every 1.77 days. The doughnut-shaped feature near the center is the site of an erupting volcano (Plume 3). The resolution is about 16 kilometers per line pair. The higher-

resolution mosaic at the right was made on the day of the spacecraft's closest approach and shows Io rotated about 120 degrees to the right, or the east. The large heart-shaped region at the lower right is the site of Plume 1, the first actively erupting volcano to be discovered on the satellite. Four months later, when *Voyager 2* arrived, the eruption had stopped. Closeup views of the same region, both active and quiescent, appear in the illustration at bottom left on page 74.

 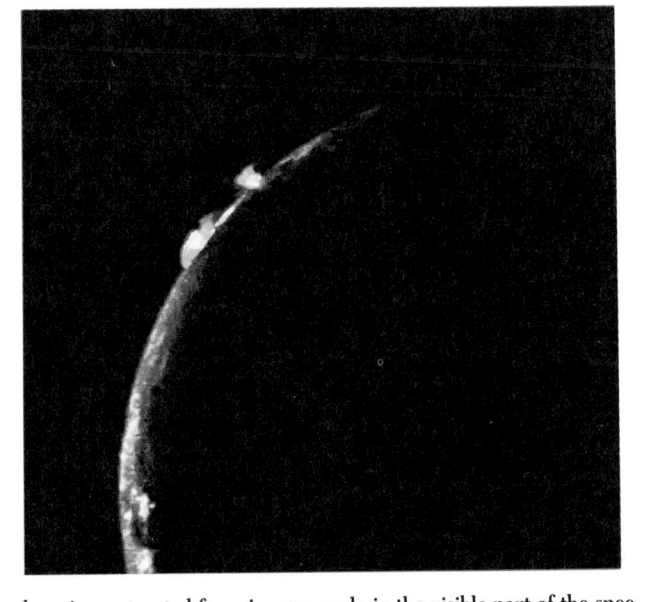

THREE MORE VOLCANIC PLUMES ON IO, of the total of eight sighted, rise to a height of about 100 kilometers in these two views. The picture at the left, made by *Voyager 1*, shows Plume 2. The one at the right, made by *Voyager 2* during a 10-hour "volcano watch" after the spacecraft's closest approach to Jupiter, shows Plume 5 and Plume 6. The image of Plume 2 is a companion to the image on the cover of the Jan. 1980 SCIENTIFIC AMERICAN. The picture of Plume 2

here is constructed from images made in the visible part of the spectrum and shows only the small central core of the plume. The picture on the cover incorporates an image made through an ultraviolet filter (passing wavelengths centered on 3,500 angstrom units) and reveals that the visible core is surrounded by a larger envelope that scatters ultraviolet radiation. In picture at right Plume 5 is the symmetrical lower plume. Plume 6, the upper one, is smaller and asymmetrical.

FIRST ACTIVE VOLCANO ON IO was discovered in this image made by *Voyager 1* on March 8, looking back 4.5 million kilometers at the satellite it had flown past three days earlier. Linda A. Morabito of the Jet Propulsion Laboratory detected the faint umbrella-shaped cloud (Plume 1) on Io's limb as she was inspecting a specially processed image. It was then recognized that the picture includes a second volcanic plume that has caught rays of the rising sun, creating a glow just inside the line of the terminator: the edge of the illuminated hemisphere.

CLOUDS OR SURFACE DEPOSITS ON IO can be seen in the vicinity of scarps in this *Voyager 1* picture of a region near the satellite's south pole. The resolution is about 16 kilometers. The white patches may be ice crystals, either aloft or on the surface, produced when liquid sulfur dioxide just under the surface breaks through fractures in the crust and explosively freezes.

observers as being possibly due to sulfur dioxide frost on the surface of the satellite. Recent spectrophotometric analysis of the Voyager images suggests that the large white regions could be about half sulfur dioxide frost and half sulfur. These findings suggest that sulfur dioxide may be abundant in Io's surface layers. If that is the case, liquid sulfur dioxide will be stable in a zone the top of which will be a few hundred meters below the surface, and it will collect in that zone.

The presence of reservoirs of liquid sulfur dioxide, much like aquifers of liquid water on the earth, may provide a simple explanation of what appear to be ice clouds issuing from fractures in Io's crust. The fluid sulfur dioxide would have easiest access to the surface along a fault or at the base of a scarp. As it reaches the surface, perhaps forced out by artesian pressures, the confining pressure drops below a critical point and the liquid explodes into an ice fog that spreads out and falls back to the surface. John F. McCauley of the U.S. Geological Survey and his colleagues have proposed that this process actually erodes material along the scarps.

The sulfur dioxide "aquifer" has also been called on to explain the much larger volcanic plumes. Bradford A. Smith of the University of Arizona and his colleagues on the Voyager imaging team have suggested that the liquid sulfur dioxide causes violent volcanic eruptions in much the same way that water does in the earth's crust. In the model proposed for Io heat, tidally generated in the silicate lithosphere, is carried upward by molten sulfur that has been in contact with the hot rock. When the molten sulfur comes in contact with the liquid sulfur dioxide, the mixture begins to rise toward the volcanic vent and rapidly expands as the pressure drops and the sulfur dioxide vaporizes. When it reaches the surface, its velocity will be between 500 and 1,000 meters per second, which is consistent with the velocities deduced for plumes rising from 100 to 300 kilometers above the surface. This model requires large segregated pools of molten sulfur, possibly large enough to justify being called oceans, below Io's surface. Some of the volcanic flows may therefore consist of almost pure sulfur, as has also been proposed by Carl Sagan of Cornell University.

At the other extreme are models proposed by Michael H. Carr and Harold Masursky of the U.S. Geological Survey in which sulfur and sulfur compounds are simply coloring agents for the lava flows of ordinary silicate volcanism, such as that on the earth. Actually diverse mechanisms may be at work in different places on the satellite. It is generally agreed that the vivid colors seen on the surface are consistent with the wide range of molecular forms of sulfur known to be stable at Io's surface temperature, which ranges between 60 and 120 degrees Kelvin.

Europa

Europa was farthest away of all the Galilean satellites at the time of *Voyager 1*'s closest approach to it (732,270 kilometers) and therefore was photographed with the lowest resolution, about 33 kilometers per line pair in the best images. This resolution is comparable to a picture of the earth's moon occupying a third of the height of an ordinary television screen. *Voyager 2*, which approached to within 204,000 kilometers, provided an eightfold improvement in resolution. The *Voyager 1* images of Europa show a body that is nearly white, with global markings and patterns that are bland and low in contrast. Its equatorial region exhibits two basic types of terrain: mottled darker regions and brighter regions, both of which are transected by a series of narrow dark stripes tens of kilometers in width extending in some instances for thousands of kilometers.

In the higher-resolution *Voyager 2* images the stripes emerge as a vast tangle of intersecting lines [*see illustration at right*]. The mottled dark terrain is resolved into a series of interlocking depressions and mesas whose typical dimension is a few kilometers. Many of the depressions may be impact craters, but their identification is uncertain.

Only three probable impact craters have been identified. They are between 18 and 25 kilometers in diameter and display widely different morphologies. One crater is fresh and bowl-shaped. Another is shallow and surrounded by a system of dark rays. The third seems to be raised on a pedestal as if the surrounding area had been etched away. In some pictures there is a hint of numerous smaller craters a few kilometers in diameter, particularly along the terminator zone (the zone at the edge of the satellite's illuminated side). The surface of Europa is at least hundreds of millions of years old and is possibly billions of years old.

The absence of substantial relief along the terminator, where it would be most visible, suggests that water, the most likely principal volatile component of the satellite, has risen from the interior to the surface and formed a thick mantle of ice that obscures topographic relief. A mantle 100 kilometers deep would certainly be thick enough to mask whatever relief could possibly exist on the silicate lithosphere.

Europa, along with all the other planetary bodies in the solar system, was presumably created by the accretion of material circling the sun four to five billion years ago. Thereafter, again like the other bodies, it continued to be subjected to an intense bombardment by me-

teorites, large and small, for perhaps another half-billion years. Its present appearance is probably a sensitive reflection of the relation between its early intense bombardment and its thermal history. If its icy crust had formed, frozen and become rigid in the period of early intense bombardment, the scars of

the impacts would be clearly visible today. Evidently its crust remained warm, soft and mobile until late enough in its history to obscure the evidence of the bombardment.

Two mechanisms have been suggested to explain how the surface temperature of the satellite could have been

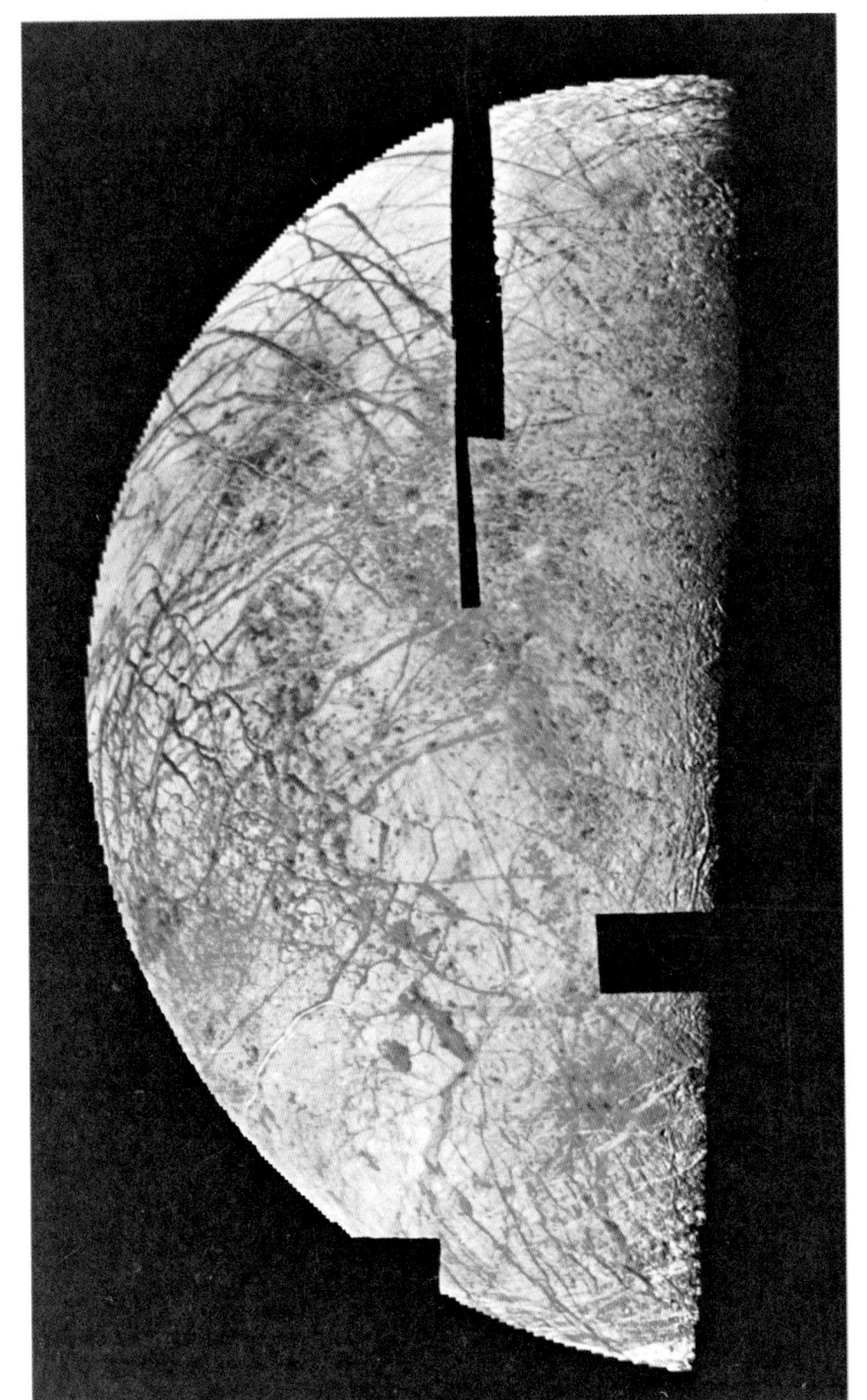

EUROPA, the next moon outward from Io, was photographed by *Voyager 2* with a maximum resolution about eight times higher than that achieved by *Voyager 1*. This mosaic of *Voyager 2* images has a maximum resolution of about 4.3 kilometers. It shows Europa crisscrossed by stripes and bands that may represent filled fractures in the satellite's icy crust. Water, in solid and liquid form, may constitute about 20 percent of Europa's mass. Satellite's albedo, or surface reflectivity, is nearly 70 percent, or 10 times the albedo of the earth's moon (and twice that of the earth). Europa circles Jupiter every 3.55 days at a distance of 600,000 kilometers.

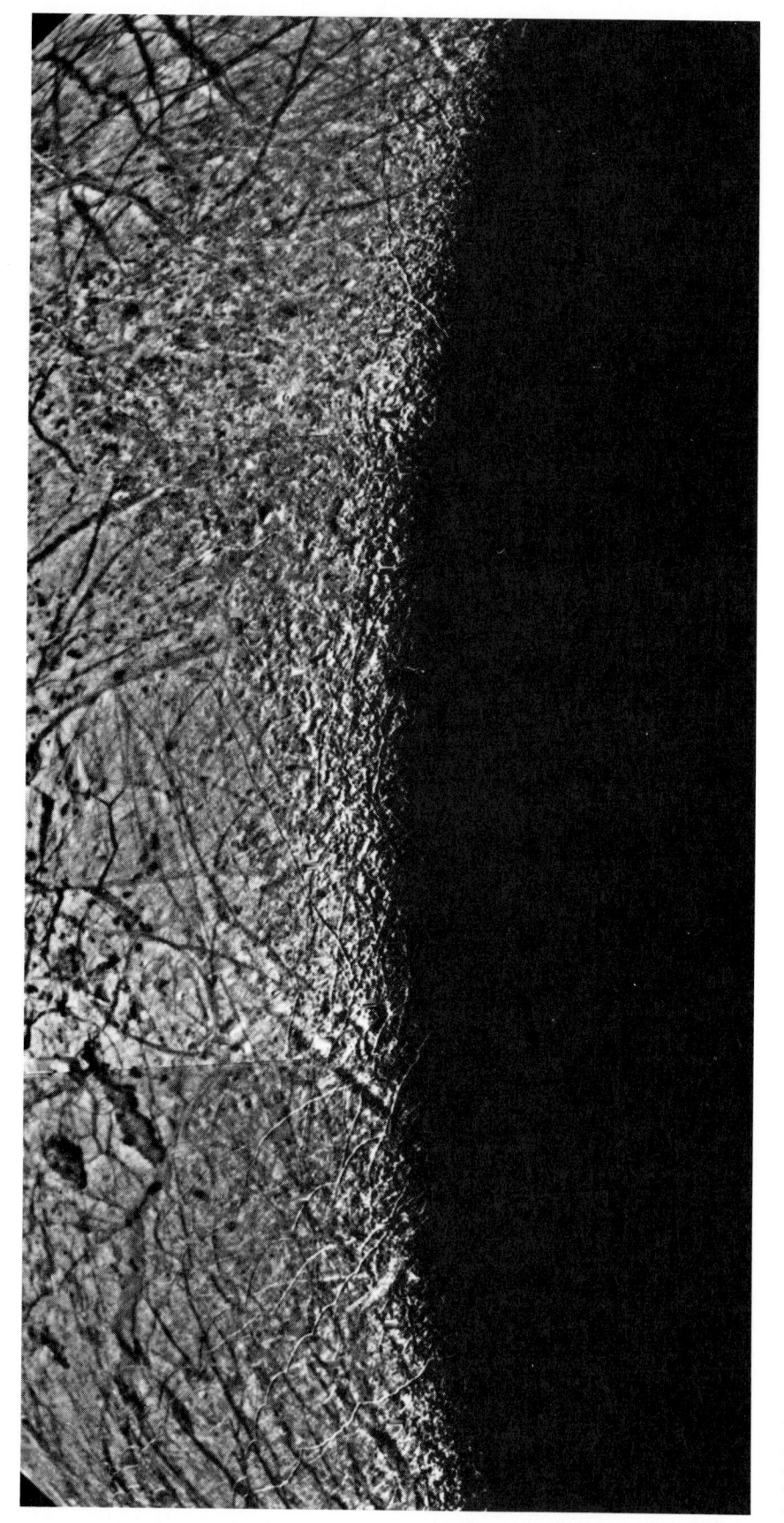

DETAIL ALONG THE TERMINATOR ON EUROPA is seen at high resolution (about 3.8 kilometers per line pair) in this mosaic of *Voyager 2* images. The low inclination of the illuminating sunlight makes it possible to estimate that the long, narrow ridges rise only about 100 meters above the average surface. The ridges are most abundant in the satellite's southern hemisphere. Much of Europa's terrain consists of interlocked depressions and mesas. Although many of the depressions could be impact craters, they are too small for positive identification.

maintained during this early period. Fraser P. Fanale and his co-workers at the Jet Propulsion Laboratory proposed a few years ago that the heat released in the decay of radioactive elements would in itself have been sufficient to keep Europa's icy mantle soft, and perhaps liquid, at depths below some tens of kilometers. Recently Cassen and his colleagues have suggested as an alternative that the same kind of tidal friction heating Io would heat Europa, although to a lesser degree. They also suggest that the dark stripes on the surface of the satellite could be filled cracks resulting from simple expansion when the oceans froze. The hypothesis may not, however, be adequate to explain the increase in area of from 10 to 15 percent implied by the width of the large dark markings. Many other models are possible.

Ganymede and Callisto

Ganymede and Callisto, the two outermost Galilean satellites, will be treated here as a pair because their bulk properties are similar and because the Voyager evidence indicates that their surfaces evolved to some extent in parallel. Although Callisto is the more distant from Jupiter, I shall take it up first because it is easier to understand and because stages in its development seem to be evident in some of Ganymede's oldest terrains.

The Jupiter-facing hemisphere of Callisto was imaged at high resolution (between 2.3 and seven kilometers per line pair) as *Voyager 1* flew past the satellite at a distance of 124,000 kilometers [*see illustration on page* 72]. The surface is almost saturated with craters, yet Callisto looks quite different from the earth's moon. For example, there is a complete absence of visible relief at Callisto's bright limb. The left half of the hemisphere is dominated by a system of concentric rings centered on a bright circular region 10 degrees north of the equator and about 600 kilometers in diameter. The rings, from 50 to 200 kilometers apart, extend out to a radius of some 1,500 kilometers.

The population of craters within the inner part of the ring system is lower by a factor of three than it is elsewhere on Callisto. The population in the outer part of the system has a density similar to that in other regions, but judging by the transection and intersection of the features some two-thirds of the craters were formed before the ring system was. Evidently preexisting craters within about 350 kilometers of the central zone were erased and craters at greater distances were preserved. Younger craters are superposed on both the rings and the central zone. The most likely explanation is that the large impact that created the rings took place early in the history of the satellite, when the crust was not rigid enough to support and retain the topographic forms normally associated with large impact basins.

The rings themselves could have been dynamically formed immediately by the impact of a large body or could have been a delayed response to the impact as the central region rebounded and the surrounding surface readjusted. Later the icy crust became rigid as it froze to substantial depths. As we shall see, a similar feature of comparable size is partially preserved on the most ancient regions of Ganymede's crust.

Pictures of Ganymede made by *Voyager 1* during its approach to the satellite show a body that remarkably resembles the earth's moon. One pattern visible at this scale that does not resemble anything seen on the moon, however, is a complex intersecting network of irregular, linear, broken and branching bright bands that crisscross the disk of the satellite. At this scale the patterns bear little resemblance to the rays of impact craters scattered elsewhere over the surface. The preliminary suggestion was that the patterns might be tectonic.

As *Voyager 1* flew closer to Ganymede two basic types of terrain were recognized: heavily cratered polygon-shaped regions up to a few tens of kilometers across surrounded by younger regions of grooved terrain. The grooved terrain was found to occupy the bright linear patterns seen at lower resolution. The grooved terrain consists of closely spaced parallel ridges and troughs, each from five to 15 kilometers wide and up to several hundred kilometers long. In some regions as many as 20 parallel grooves and ridges can be counted.

The density of the craters on the grooved terrain of Ganymede is extremely variable, ranging from a density equivalent to that found in the cratered terrain to a density lower by a factor of 10. The implication is that the formation of the grooved terrain began early in the satellite's history and continued for a long time through the period of intense bombardment. Most of the grooved terrain is a mosaic of discrete systems, with the grooves of one system ending abruptly at the boundary of an adjacent system. In some instances, however, several systems of grooves transect one another. In such instances the older system is not offset by the younger, implying that the process that forms the grooves does not involve lateral motion. It has been suggested that the grooves are formed by a process similar to the one that creates intrusive dikes in the earth's crust: fluid material might be injected into faults and then freeze.

The encounters of the two Voyagers with Ganymede were highly complementary. *Voyager 1* viewed the hemisphere of the satellite facing Jupiter; *Voyager 2* encountered the satellite before the spacecraft's closest approach to

GANYMEDE is the largest of Jupiter's moons. This picture, which was made by *Voyager 2* at a distance of 1,217,000 kilometers, shows the hemisphere of the satellite that always faces away from the planet. The hemisphere is dominated by an immense dark area, the largest remnant of an ancient crust riddled by the impact of meteorites. Early in its history Ganymede may have been as densely cratered as Callisto is today. Ganymede is only two-thirds as dense as Europa and therefore may have a higher fraction of water, perhaps as much as 50 percent.

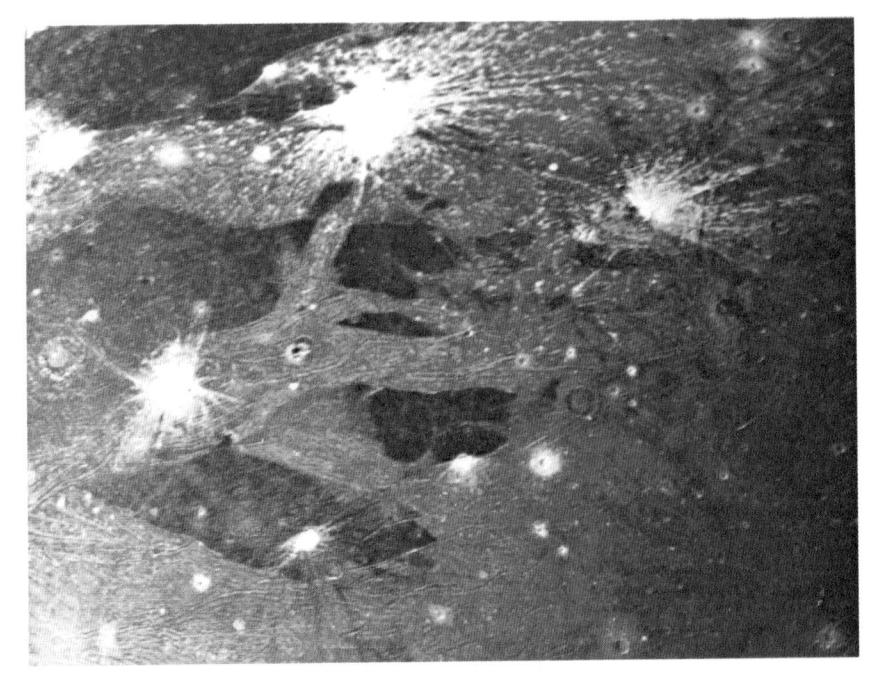

GANYMEDE'S JUPITER-FACING HEMISPHERE was photographed at high resolution by *Voyager 1*. The area shown is about 1,600 kilometers across and has a maximum resolution of about four kilometers per line pair. Ganymede's crust presumably consists mostly of ice, which is darkest in oldest regions. Around rayed craters ice may be white because it is cleaner.

Jupiter and therefore got high-resolution coverage of the opposite hemisphere [*see top illustration on page 81*]. The most striking feature in the *Voyager 2* view of Ganymede is a large circular region of ancient dark cratered terrain covering about a third of the entire hemisphere. It is the only feature that can easily be identified with telescopes on the earth. Traversing the dark region is a series of giant bright streaks that are slightly curved and parallel. In the higher-resolution images they bear a striking resemblance to the ring structures on Callisto. Although studies have shown that the streaks on Ganymede form concentric circles, there is no evidence at their center of symmetry of a bright impact-disrupted zone like the one on Callisto. Evidently the formation of younger terrains, primarily grooved ones, has erased the record of an ancient impact site.

Although the ring systems on Callisto and Ganymede exhibit strong similarities, where the rings on Callisto take the form of flat-topped ridges those on Ganymede are furrows. The furrows suggest what the ridges on Callisto looked like before they were filled. Such differences may reflect variations in the detailed history of the two bodies, including differences in the strength of the early crust, differences in the chronologies of cooling and freezing and differences in the

availability of fluid materials capable of being extruded.

Another important observation is the wide variation in the appearance of impact craters on Ganymede. Some of the craters look fresh and young. At the other extreme are ancient circular forms that are hardly more than "ghosts" or stains on the ancient crust. Evidently the brighter ghost craters were formed when Ganymede's crust was too warm and soft to hold a record of the impact that made them. Close examination indicates that the younger or smaller the crater is, the better preserved its relief is. Eugene M. Shoemaker and his colleagues in the U.S. Geological Survey have concluded that the cratering record in ancient terrains on Ganymede and Callisto is also a record of the thermal gradient and the strength of the upper crust as a function of time as the two bodies began to cool.

The discovery by *Voyager 2* of a large fresh impact basin near the south pole of Ganymede is further evidence of major changes in the strength and rigidity of the satellite's crust. The basin resembles some of the major impact basins of the inner solar system. In high-resolution views one can see that the flat-floored basin is surrounded by blocky massifs and a blanket of ejecta, indicating that by the time the basin had formed the crust had hardened enough to support

large-scale mountainous relief. The fact that numerous large craters are superposed on the basin suggests it was formed in the first billion years after the accretion of Ganymede, when the satellite was still being bombarded by bodies of substantial size. Evidently the crust was cooling rapidly and freezing during the first few hundred million years of the satellite's existence.

By comparing Ganymede and Callisto it is possible to reconstruct the evolutionary history of Ganymede in broad strokes. One can imagine that the ancient surface of the satellite was soft, icy and dark, resembling the present surface of Callisto. The huge array of slightly curved parallel streaks in the northern hemisphere was created when a large meteorite punctured the soft crust and set up oscillations and strains that fractured the crust in an enormous pattern of concentric rings. Water filled the fractures and froze. Later polar caps were formed, and grooved terrain slowly began to develop and grow. As sections of the crust slid past each other faults at an angle to the lines of slippage gave rise to offsets in the lines. Then the crust became rigid enough to preserve a lasting record of meteorite impacts. The young impact basin near the south pole and a large bright-rayed crater near the equator were the last large-scale features created in the hemisphere facing away from Jupiter that was viewed by *Voyager 2*. In the several billion years since these early events in the evolution of Ganymede the satellite has evidently remained quiescent.

History in the Craters

One of the major objectives of planetary science is to compare the geological histories of the terrestrial planets and satellites in the solar system as they are revealed by the detailed images returned by spacecraft. The principal means of establishing the relative time scales of these histories from images is to determine the record of impact cratering on the simple premise that the older the surface is, the more craters there are on it. Establishing absolute time scales from the cratering record is much more difficult because it requires a knowledge of the cratering rates for each of the bodies.

There is now ample evidence on how the rate of cratering differed from one part of the solar system to another. Models of the rates at which craters were made by the impact of asteroids (and to a lesser extent of comets), together with observations of the crater populations on different bodies, suggest that the average rates over the past few billion years were similar for Mercury, Mars, the earth and the moon within a factor of perhaps two.

Shoemaker has suggested, however,

GROOVED AND TWISTED TERRAIN is a distinguishing feature of Ganymede. This *Voyager 1* view along the terminator from a distance of 135,000 kilometers has a resolution of about 2.4 kilometers. The closely spaced parallel ridges and troughs are usually from five to 15 kilometers across. Younger systems of grooves can be seen transecting older systems. In the area shown here the grooved terrain has totally replaced heavily cratered ancient dark crust.

that over the past few billion years the cratering rates for Jupiter's satellites may have been substantially lower than those for the inner solar system, perhaps by a factor of between 10 and 100. His reasoning is as follows. Although asteroids are deflected outward from the inner solar system to the region of Jupiter's orbit, they are probably soon deflected again by Jupiter's powerful gravitational field to other parts of the solar system. Therefore they have only a small probability of hitting Jupiter or its satellites. That being the case, the flux of smaller objects in the Jovian system is probably dominated by periodic comets, which are less abundant than asteroids by a factor of at least 10 and perhaps as much as 100.

Another factor that influences the cratering rates of the satellites is Jupiter's great mass, which dramatically increases the velocity of incoming objects. Thus an impacting body of a given size would tend to produce a larger crater on a satellite near Jupiter than it would on one farther away. This enhancement of velocity is negligible for Callisto, but it is about a factor of two for Ganymede, three for Europa and five for Io.

When crater size is plotted against impact-frequency distribution for the Galilean satellites and for representative bodies of the inner solar system, one obtains curves for Callisto and Ganymede that are rather like those for the earth's moon and Mars [see illustration at right]. In contrast Europa (on the basis of only three probable craters) and the earth have fewer craters of a given diameter than the other bodies by a factor of between 10 and 50. Although Europa may have many small craters less than 15 kilometers in diameter, direct evidence for them is lacking. Io, as we have seen, has no craters at all down to the limit of Voyager image resolution. The low cratering value for the earth undoubtedly reflects the difficulty of identifying small craters and the obliterating effects of geological processes.

The following general conclusions can be reached. The heavily cratered terrains on Ganymede and Callisto, comparable to the heavily cratered highlands on the earth's moon, Mars and Mercury, must date back to the period of torrential bombardment some four billion years ago. Evidently the grooved terrain on Ganymede started to form before the early intense bombardment ended.

A lower limit on the recent cratering rates can be derived as follows. First, the surface of Europa certainly cannot be older than about four billion years or it would still bear the scars of large ancient craters formed during the early torrential bombardment. Three identified craters in four billion years sets a lower limit for the flux of crater-making objects in the vicinity of Europa at

about a tenth of the flux in the vicinity of the earth's moon. This is consistent with Shoemaker's estimate of the flux in the vicinity of Europa and could imply that the surface of Europa is billions of years old. At the other end of the range of possibilities, the surface of Europa could not be younger than perhaps

100 million years or the flux in the vicinity of Europa would have to be much greater than that in the vicinity of the moon. As for Io, regardless of flux estimates the total absence of craters simply means that it has the youngest and most dynamic surface yet observed in the solar system.

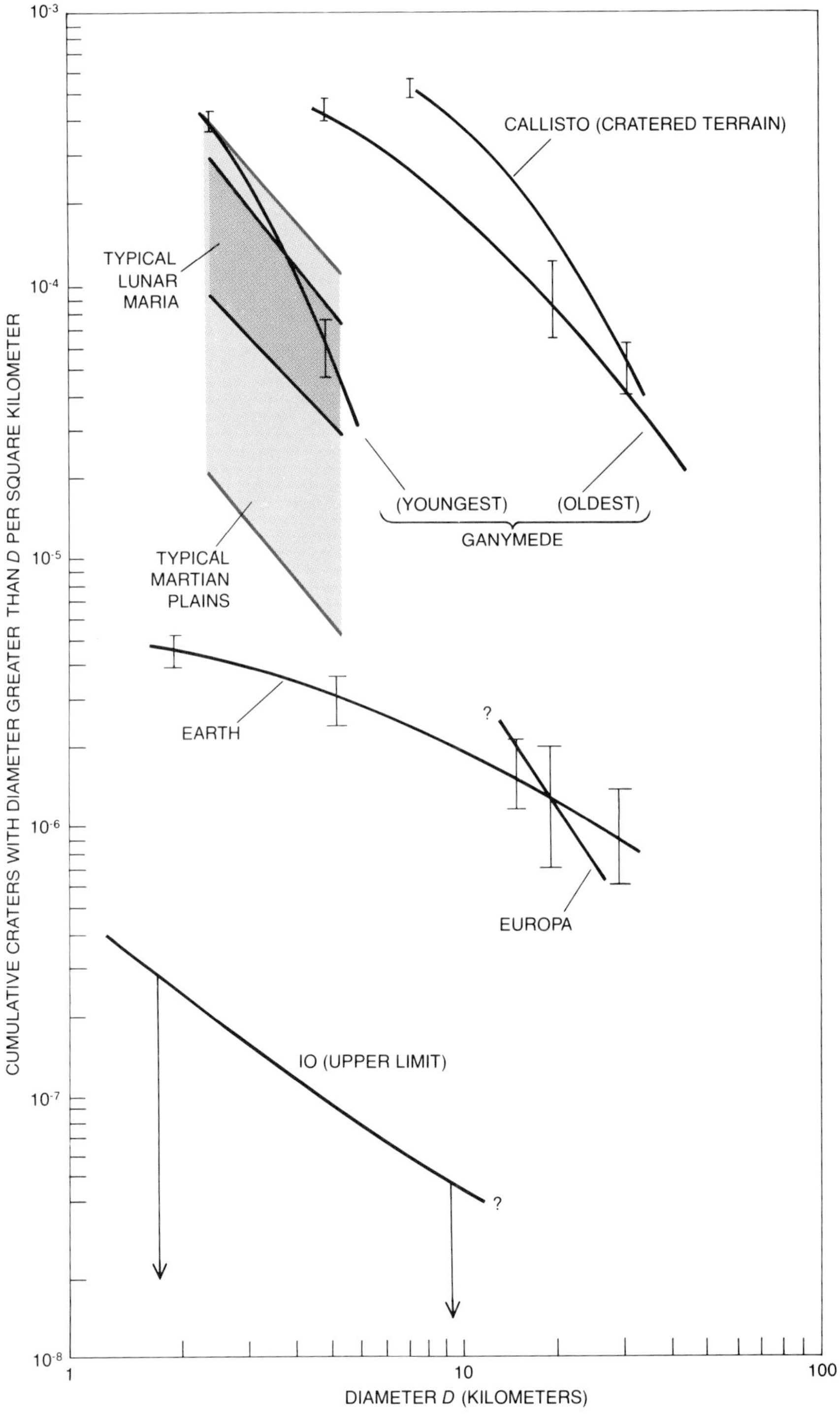

SIZE AND FREQUENCY OF THE IMPACT CRATERS on Europa, Ganymede and Callisto fall within the range measured on the earth, the earth's moon and Mars. Io, however, has no impact craters down to the resolution of the Voyager images of the satellite: about one kilometer. The curve for Europa is highly uncertain since only three craters can be confidently identified in the Voyager pictures. Estimates for impact craters on the earth were made by Richard Grieve and Michael R. Dence of Canadian Department of Energy, Mines and Resources.

8 Titan

by Tobias Owen
February, 1982

The largest moon of Saturn is the only moon in the solar system with a substantial atmosphere. The chemistry of the atmosphere may resemble that of the earth's atmosphere before life arose

On November 12, 1980, the spacecraft *Voyager 1* passed within 7,000 kilometers of Titan, the largest moon of Saturn. It was the closest encounter between *Voyager 1* or *Voyager 2* and any planet or moon, but it was made at a cost. The encounter meant that *Voyager 1* could not be redirected by the gravitation of Saturn so that it would travel onward through the solar system to pass near Uranus and Neptune. The sacrifice seemed appropriate; Titan was known to be the only moon in the solar system that has a substantial atmosphere. Moreover, its reddish color, which is unique among Saturn's moons, suggested that the chemistry of Titan's atmosphere might be producing colored compounds. Because of the closeness of the encounter the instruments on board the spacecraft would be able to perform at their best.

The results of the encounter show how wise the sacrifice was. Titan turns out to be the only known body in the solar system besides the earth whose surface is at least partially covered by a liquid. On Titan the liquid is methane. Moreover, the Voyager instruments showed that the atmosphere of Titan is denser than the atmosphere of the earth. This denser atmosphere has retained conditions much like those that probably existed on all the planets soon after they formed. Specifically, the atmosphere of Titan has carbon, nitrogen and hydrogen but lacks molecular oxygen. Under these conditions the chemical reactions proceeding in Titan's atmosphere today may well be giving rise to some of the organic molecules that are thought to have been precursors to life on the earth.

The Presence of an Atmosphere

Christiaan Huygens discovered Titan in the spring of 1655, the year in which he also proposed that Saturn has rings. The body was named almost two centuries later, when Sir John Herschel assigned names to the seven moons of Saturn that were known at the time. The name Titan was well chosen. Her-schel knew only that Titan was the brightest moon of Saturn; since then it has proved to be also the largest. Indeed, it is larger than the planet Mercury. For a time it was thought to be the largest moon in the solar system. Measurements made by *Voyager 1* show that it does not quite hold the record. The earlier measurements had been inflated by the thickness of Titan's atmosphere. The solid body of Titan has a radius of 2,575 kilometers. Jupiter's moon Ganymede is larger: its radius is 2,640 kilometers.

The first hint that Titan has an atmosphere came from observations of the body that the Catalan astronomer José Comas Solá published in 1908. Solá reported that the tiny disk of Titan he could see through his telescope was darker at its limb, or periphery, than it was at its center. The reason, he proposed, was the presence of an atmosphere. Specifically, the sunlight reflected toward the earth by Titan's limb must pass through more of Titan's atmosphere than the sunlight reflected by the center of Titan's disk. Thus the light from the limb is attenuated in greater measure by absorption in Titan's atmosphere.

It is hard to know whether Solá really observed that the limb of Titan was darkened. The descriptions he gave of patchy clouds on the giant moons of Jupiter are known to be mistaken. Nevertheless, his observations seem to have led Sir James Jeans to include Titan and the giant moons of Jupiter in his theoretical study of the escape of atmospheres from the bodies of the solar system. In 1916 Jeans concluded that Titan has probably retained an atmosphere in spite of its small size and weak gravity compared with, say, the earth because of its low temperature. Titan's distance from the sun combined with any reasonable estimate of its reflectivity (and loss of solar heating on that account) leads to the prediction that its surface and its atmosphere should have a temperature of between 60 and 100 degrees Kelvin. For such a range Jeans's work shows that a gaseous substance whose molecular weight is 16 or greater should not have escaped from Titan over the history of the solar system.

Several substances satisfy Jeans's limit on weight. One of them is ammonia (NH_3), whose molecular weight is 17. In the 1930's Rupert Wildt of the University of Göttingen identified it as a component of the atmosphere of Jupiter. Wildt had found that the spectrum of the sunlight reflected from Jupiter in the infrared showed the absorption of the radiation at wavelengths characteristic of ammonia molecules. By a similar method Theodore Dunham, Jr., of the Mount Wilson Observatory detected ammonia on Saturn. At the temperature assumed for Titan, however, ammonia would be a solid; it cannot be a substantial part of an atmosphere. Other substances that satisfy the limit are argon, neon and molecular nitrogen (N_2). All of them would have had an appreciable concentration in the mixture of gases and dust that condensed to form the solar system. The problem is that they are hard to detect spectroscopically. None of them absorbs much radiation in the infrared.

Still another substance that satisfies the limit is methane (CH_4), whose molecular weight is 16. Unlike argon, neon and molecular nitrogen it has a strong set of absorption bands in the infrared, and unlike ammonia it is gaseous at the temperature predicted for Titan. In 1932 Wildt identified methane in spectra of Jupiter, Saturn, Uranus and Neptune. Then in 1944 Gerard P. Kuiper of the University of Chicago identified it in the spectrum of Titan. His discovery constituted the first strong evidence that Titan indeed has an atmosphere. By comparing the spectrum of Titan with laboratory spectra of methane at low pressures Kuiper deduced that the absorption of sunlight by the gas along a vertical path through Titan's atmosphere is equivalent to the absorption of such radiation by a column of methane 200 meters long at a pressure of one earth atmosphere and a temperature of 273 degrees K. (Such conditions are called standard temperature and pressure, or STP.) For the purpose of comparison a vertical column through the earth's atmosphere amounts to a column eight kilometers long at STP.

Some problems with the early understanding of Titan's atmosphere began to materialize in 1965, when Frank J. Low of the University of Arizona deduced from the brightness of the radiation Titan emits at an infrared wavelength of 10 micrometers that the temperature of the body is 165 degrees K., a temperature nearly twice as high as the temperature one would attribute to simple solar heating of Titan's surface and lower atmosphere. Low's finding went largely unnoticed for seven years. Then a number of investigators began to find other surprises. For one thing, the calculations of Titan's temperature based on measurements of its brightness at various infrared wavelengths failed to yield a consistent value. Titan's "brightness" at radio wavelengths also yielded inconsistencies. The radio brightness in one observation implied a surface temperature of 200 degrees K.

Further still, the light reflected from Titan at the small angles that prevail between the sun, Titan and the earth turned out to have a positive polarization: the vector representing the maximum strength of the electric component of the electromagnetic field of the light was perpendicular to the plane in which the angle lay. This finding, made independently by Joseph F. Veverka of Cornell University and Benjamin H. Zellner of the University of Arizona, suggested that the light reflected from Titan came not from the solid surface of the body through a transparent atmosphere but from a deep, cloud-filled atmosphere. Mars, for example, has only a thin atmosphere, and the light it reflects toward the earth at small angles shows negative polarization: the electric-vec-

ATMOSPHERE OF TITAN is apparent in an image of the night side of the body made last August 25 by the spacecraft *Voyager 2.* The orange crescent forming the left part of Titan's limb represents the reflection of sunlight by an aerosol: a layer of particles suspend- **ed in Titan's atmosphere some 200 kilometers above the surface. The blue halo surrounding both the crescent and the unlit part of Titan's limb represents the scattering of sunlight at large angles by particles of haze suspended as high as 300 kilometers above the surface.**

tor maximum is in the plane in which the angle lies.

Meanwhile Laurence M. Trafton of the University of Texas at Austin had surmised that the amount of methane on Titan must be considerably greater than the amount deduced by Kuiper or else some other gas must also be present in quantity. With the aid of an infrared image intensifier that had just become available Trafton had found that a methane absorption band near a wavelength of one micrometer in the spectrum of Titan was unexpectedly strong.

The strength of the band could be due to methane in surprising abundance. Alternatively the collisions between methane molecules and molecules of an undetected gas could be perturbing the vibrational states of the methane molecules, so that the methane would be absorbing infrared radiation over a broad range of wavelengths. In either case the strength of the absorption band was related to both the abundance of the methane and the local atmospheric pressure.

Trafton further discovered that the stronger methane absorption bands in

Titan's infrared spectrum have an appearance different from that of the same bands in spectra of Jupiter and Saturn. Titan's bands are both shallower (less intense) and broader. This finding suggested to Trafton that small particles are suspended in Titan's atmosphere in numbers and at altitudes different from those of the particles composing the hazes observed on Jupiter and Saturn. The particles scatter sunlight toward the earth; thus they brighten the absorption bands. They also scatter sunlight at angles in Titan's atmosphere. The in-

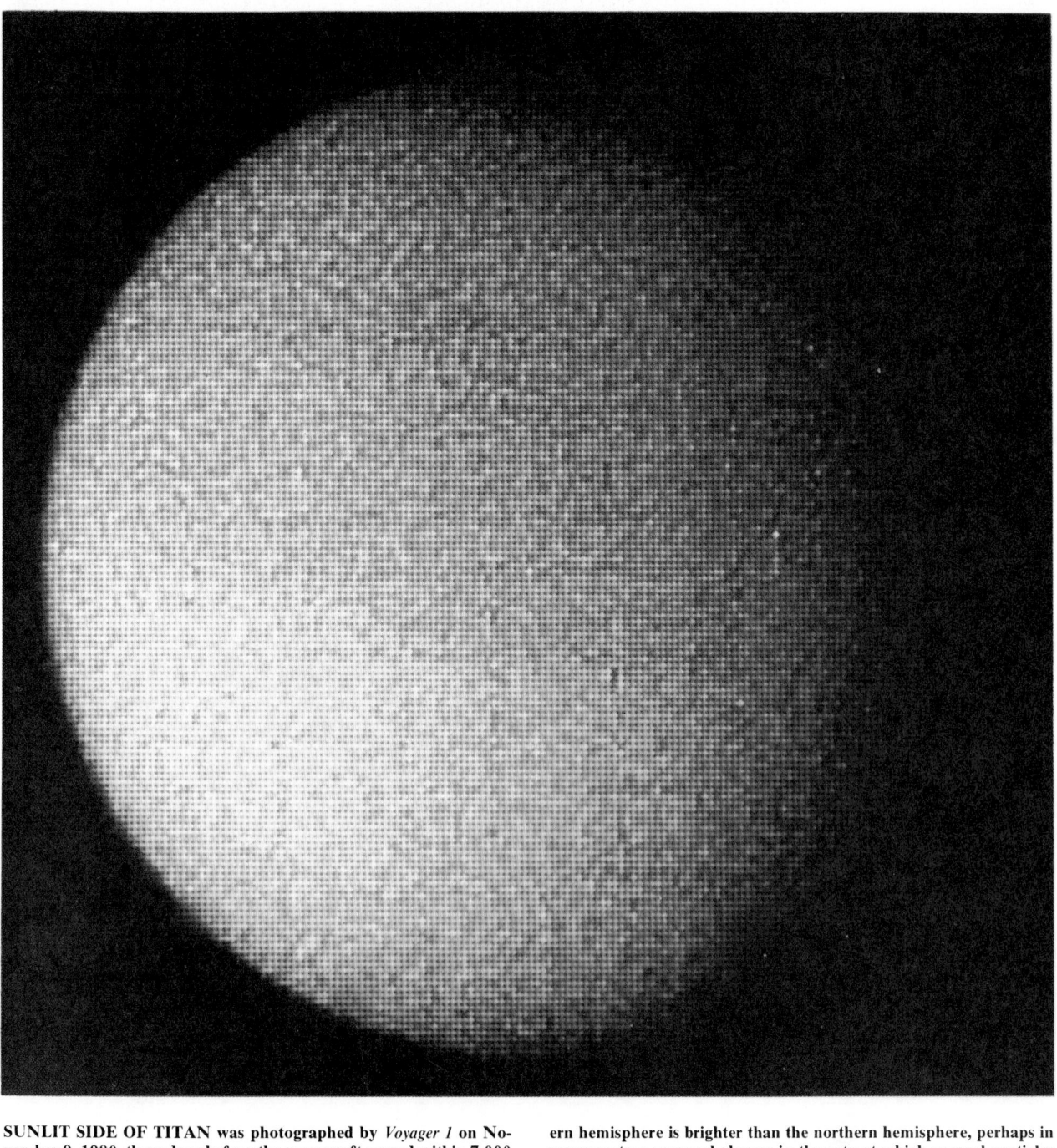

SUNLIT SIDE OF TITAN was photographed by *Voyager 1* on November 9, 1980, three days before the spacecraft passed within 7,000 kilometers of the body. In the resulting image the surface of Titan is concealed by the unbroken opacity of the aerosol layer. The south-ern hemisphere is brighter than the northern hemisphere, perhaps in response to a seasonal change in the rate at which aerosol particles are produced. It is spring in the northern hemisphere; thus the south-ern hemisphere has just passed through a summer seven years long.

creased path length of a photon (a quantum of light) scattered in this way makes it more likely that the photon will encounter a methane molecule and be absorbed by it. The likelihood increases even for photons whose wavelength is somewhat different from the wavelength that defines the center of an absorption band. (According to quantum theory, such absorptions are unlikely but not impossible.) In this way the bands are broadened.

Models for Titan

At the State University of New York at Stony Brook, Barry L. Lutz, Robert D. Cess and I turned to a large set of laboratory spectra of methane made by Lutz in an effort to improve our understanding of the outer planets and Titan. We found that the absorption bands Wildt and Kuiper had studied are insensitive to pressure. Each of the bands consists of narrow absorption lines spaced so that any broadening of the individual lines by an increase in the pressure of the methane has no effect on the band overall. From the strength of the bands we deduced the abundance of methane on Titan. It was equivalent to a column of methane about 120 meters long at STP. Then from the relation between abundance and pressure that Trafton had proposed we deduced the atmospheric pressure that broadens the methane bands on Titan. (In this analysis we ignored the effects of the light scattered by particles.)

The result surprised us. We found that the pressure of Titan's atmosphere at the base of the zone of the atmosphere that reflects light is nearly 400 millibars, or almost half the pressure at sea level on the earth, which is approximately one bar. The methane by itself would give rise to a pressure of only one millibar. Trafton's second alternative seemed to be the right one: a gas other than methane is present in large quantity on Titan.

As we attempted to deduce the pressure of Titan's atmosphere other workers were attempting to account for the contradictory measurements of the temperature of the body. The first idea had been that the apparent high temperature of Titan, such as the 165 degrees K. reported by Low, results from the "greenhouse" effect. This concept assumes that the atmosphere of Titan is transparent to visible light but opaque to the infrared. Hence sunlight penetrates to the surface and heats it; then the surface gives off infrared radiation, which is trapped in the lower atmosphere. The concept had been compromised by the accumulating contradictory measurements, which suggested that the atmosphere has an unusual structure in which the upper layers are warmer than the surface. Thus the greenhouse concept had given way to a model championed

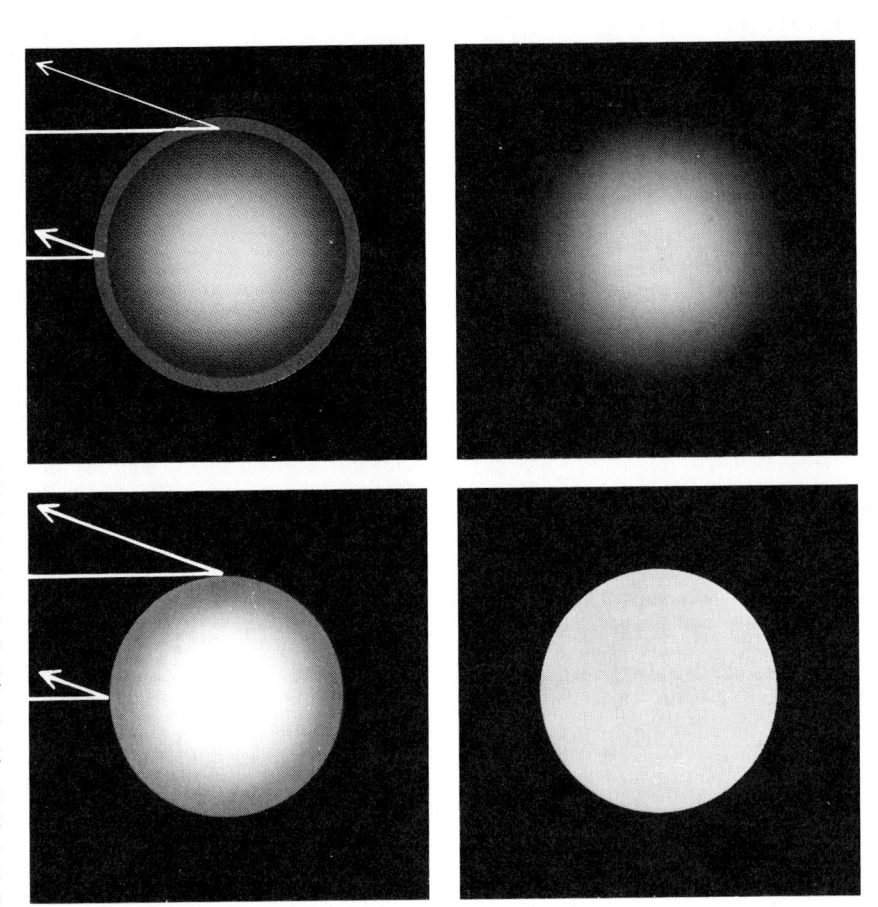

LIMB DARKENING causes the image of a moon or planet that has an atmosphere (*top*) to be darker at its limb, or periphery, than at its center. The darkening arises because the light reflected from the limb traverses a longer path through the body's atmosphere than the light reflected from the center. It is therefore attenuated by atmospheric absorption to a greater degree. In contrast the image of a planet or moon without an atmosphere (*bottom*) is a disk of more or less uniform brightness. In 1908 the Catalan astronomer José Comas Solá suggested that Titan must have an atmosphere because its image in his telescope showed darkening toward the limb.

by Robert E. Danielson and John J. Caldwell of Princeton University.

In that model the infrared brightness of Titan is ascribed to the emission of infrared radiation by an inversion in the atmosphere: a layer in which the temperature of the atmosphere increases with altitude instead of decreasing. Specifically, in the model the atmosphere is heated because sunlight is absorbed by particles that make up an aerosol (a suspension of particles) high in the atmosphere. The infrared radiation detected on the earth is emitted by methane and also by gases such as ethane (C_2H_6) and acetylene (C_2H_2), the products of light-induced chemical reactions in which methane molecules are destroyed by ultraviolet radiation from the sun. According to a later elaboration of the model by Caldwell, the atmosphere of Titan is 90 percent methane. Its surface pressure is 20 millibars and its surface temperature is 86 degrees K., a value in agreement with the prediction that would be made on the basis of simple solar heating of a body at Titan's distance from the sun. The evidence favoring a higher surface pressure than Caldwell contemplated could be rejected on

the grounds that the scattering of sunlight by the aerosol particles could give the methane absorption bands the same appearance they would get from the pressure of a gas other than methane.

An alternative to this model was one proposed by Donald M. Hunten of the University of Arizona. Hunten pointed out that the light-induced dissociation of ammonia molecules in Titan's atmosphere could lead to the accumulation of molecular nitrogen; the hydrogen liberated by the dissociation would rapidly escape from the body. Since nitrogen is transparent to visible light and to infrared radiation it could not be detected spectroscopically from the earth. Hunten demonstrated, however, that if Titan's atmosphere includes enough molecular nitrogen to contribute 20 bars to the pressure at the surface, the increased pressure could lead to the increased absorption of infrared by nitrogen itself through collision-induced absorption. In this way a greenhouse effect could heat the surface of Titan to the temperature of 200 degrees K. inferred from one of the existing radio measurements.

Hunten's model included the aerosol layer and the corresponding temperature inversion that Danielson and Caldwell's model had placed in the upper atmosphere. Other than that the models disagreed disturbingly. To summarize, Danielson and Caldwell proposed that the surface of Titan has a temperature of 86 degrees K. and that the atmosphere is 90 percent methane; Hunten proposed a surface temperature of 200 degrees and an atmosphere of 90 percent nitrogen. Worst of all, Danielson and Caldwell proposed a surface pressure of 20 millibars; Hunten proposed 20 bars, a figure 1,000 times greater.

It seemed possible that this worst discrepancy could be reduced, and perhaps even eliminated, by better measurements of the brightness of Titan at radio wavelengths. The point of making such measurements is that most of the predicted constituents of an atmosphere in the outer solar system, including methane and nitrogen, are transparent at radio wavelengths; therefore the measurements would probably represent the thermal emission from Titan's surface and not a layer in the atmosphere. A problem encountered in such measurements is that Titan's weak emission is swamped by the radiation emitted by Saturn. To diminish the problem Walter J. Jaffe of the National Radio Astronomy Observatory, working with Caldwell and me, employed the Very Large Array of radio telescopes in New Mexico. We made observations of Titan at the radio wavelengths of 1.3, two and six centimeters. The Very Large Array has sufficient angular resolution to separate Titan's emission from Saturn's. We found that Titan has a surface temperature of 87 degrees K. plus or minus nine degrees. Our finding allowed Danielson and Caldwell's model. It also allowed Hunten's model if that model is modified to have nitrogen contribute a maximum of two bars to the surface pressure. We had reduced the discrepancy, but only by a factor of 10.

Exploration by Voyager

That was how matters stood in the fall of 1980 as *Voyager 1* approached the Saturn system. Titan was known to be a large moon with an atmosphere at least three times denser than that of Mars (where the average surface pressure is seven millibars). The postulated presence of ethane, acetylene and a high-altitude aerosol suggested an active photochemistry. Methane had been detected spectroscopically; the presence of nitrogen had been proposed. Thomas W. Scattergood and I suggested the red color of Titan was indirect evidence that nitrogen was indeed present. Scattergood had attempted to make colored compounds by bombarding various mixtures of gases with energetic protons. His intent was to simulate the bombardment of Titan's atmosphere by charged subatomic particles trapped in the magnetic field surrounding Saturn. His efforts were unsuccessful if the gas was methane alone. If the gas was a mixture of methane and nitrogen, however, the bombardment led to the production of a reddish-brown material. But how much nitrogen was present on Titan? And what other gases were there?

The first results of the arrival of *Voyager 1* near Titan were images of the body. They were rather disappointing. Some investigators had hoped to see breaks in the aerosol layer that would allow a glimpse of Titan's surface. Instead the images showed a moon that resembled a fuzzy, seamless tennis ball. The aerosol was ubiquitous and opaque. The only markings visible in it were a dark north-polar hood and an abrupt change in reflectivity at the equator, so that the southern hemisphere was distinctly brighter than the northern. Titan was also shown to be surrounded by a high-altitude layer of haze about 100 kilometers above the top of the aerosol layer.

The difference in reflectivity between the northern and southern hemispheres has received two tentative explanations. On a hypothesis developed primarily by Lawrence A. Sromovsky and Verner E. Suomi of the University of Wisconsin at Madison the difference is a manifestation of the seasonal change in the solar heating of the aerosol layer caused by the 26-degree inclination of Titan's axis of rotation with respect to the plane of the solar system. If the hypothesis is correct, the southern hemisphere should alternate with the northern hemisphere in being the brighter one. Saturn and its moons take 30 years to circle the sun; hence the alternation should come once every 15 years.

On another hypothesis, developed by G. Wesley Lockwood of the Lowell Observatory, the difference results from a modulation in the production of aerosol particles due to the differing rates of arrival of the high-energy subatomic particles of the solar "wind." Over the past eight years Lockwood has been recording small changes in the net brightness of Titan. For much of that time the number of spots on the sun has been increasing, and with this increase the solar wind has been intensifying. Continued observations of Titan throughout the next few years, in the waning part of the sunspot cycle, should help to test the two hypotheses. They both may be correct.

After the images of Titan came other data from *Voyager 1*. For one thing, the ultraviolet spectrometer on board the spacecraft revealed to a group led by Lyle Broadfoot of the University of Southern California the presence of nitrogen: the instrument detected peaks in the ultraviolet spectrum of Titan due to the emission of ultraviolet radiation by nitrogen molecules, ionized nitrogen atoms and un-ionized nitrogen atoms. The spectrum gave no suggestion of carbon monoxide, argon or neon (other substances that emit in the ultraviolet). On the other hand, the presence of methane and other hydrocarbons was suggested as the spectrometer monitored the absorption of the light from a star by Titan's atmosphere. It was in essence an occultation experiment in which the limb of Titan, and therefore Titan's atmosphere, intervened between the star and the spectrometer.

In a second occultation experiment the spacecraft itself was the source of the radiation. Here the radio signals beamed toward the earth by *Voyager 1*'s transmitters were attenuated by refraction in Titan's atmosphere. The attenuation grew as the spacecraft disappeared behind the limb of Titan and the density of gas traversed by the beam increased. The result of the experiment was a profile of density with respect to altitude in Titan's atmosphere. The profile of density, in turn, yields a profile of $T/\bar{\mu}$ with respect to altitude, where T is temperature and $\bar{\mu}$ is the mean molecular weight of the atmosphere. The experiment had been designed by G. Leonard Tyler of Stanford University to allow for either of the two prevailing models of Titan's atmosphere: Danielson and Caldwell's or Hunten's.

In the end it favored Hunten's. In particular, Von R. Eshleman of Stan-

DISTINCT LAYER formed by Titan's high-altitude haze particles is shown in a photograph of Titan's south pole made by *Voyager 2*. The haze layer is concentric with the brightest feature in the photograph: the southern cusp of the sunlit crescent of Titan's opaque aerosol layer.

ford pointed out that the profile of $T/\bar{\mu}$ produced by the experiment closely matches the predictions of Hunten's model for an atmosphere rich in nitrogen except that the profile implies a surface pressure lower than Hunten's original value of 20 bars. The group responsible for the occultation experiment is continuing to analyze its data. Gunnar F. Lindal of the Jet Propulsion Laboratory of the California Institute of Technology has given the group's most recent report of a surface pressure of 1.5 bars (plus or minus .1) and a surface temperature close to 94 degrees K. Lindal has also presented a value of 2,575 kilometers (plus or minus two) for the radius of Titan as derived from the occultation.

The structure of Titan's atmosphere was also analyzed by an infrared spectrometer group led by Rudolf A. Hanel of the Goddard Space Flight Center of the National Aeronautics and Space Administration. The best mutual fit of the occultation data and the infrared data indicates that the mean molecular weight of Titan's atmosphere is 28.6; hence the atmosphere must include an appreciable amount of a gas heavier than nitrogen. (The molecular weight of nitrogen is 28.0.) Robert E. Samuelson of the Goddard Space Flight Center suggests the gas is argon. He and his colleagues reason that argon is relatively abundant in the universe and that it is gaseous at Titan's temperature. Moreover, argon (like nitrogen) is transparent in the visible and the near infrared, so that it escapes detection by infrared spectroscopy.

The fact that argon was not detected by the ultraviolet spectrometer on *Voyager 1* means only that its abundance in the upper atmosphere of Titan is less than 6 percent. In the upper atmosphere one expects the lighter gases to dominate. Hence the limit of 6 percent in the upper atmosphere of Titan does not rule out the net abundance of about 12 percent required to give the atmosphere a mean molecular weight of 28.6. By setting $\bar{\mu}$ equal to 28.6 in the profile of $T/\bar{\mu}$ one finds a surface temperature of 95 degrees K., plus or minus two degrees.

The data transmitted to the earth from the vicinity of Titan by *Voyager 1* have thus revealed a moon in the outer solar system whose atmosphere, like the earth's, is rich in nitrogen. Furthermore, the data show that this body's atmosphere has a surface pressure greater than that of the earth's. The surface pressure of a body's atmosphere represents both the quantity of gas in the atmosphere and the degree to which it is compressed by the gravitation of the body. The gravity at the surface of Titan is only .14 times as strong as the gravity at the surface of the earth. Remarkably, therefore, the surface pressure on Titan (1.5 bars, according to the Voyager data) means that the atmosphere of Ti-

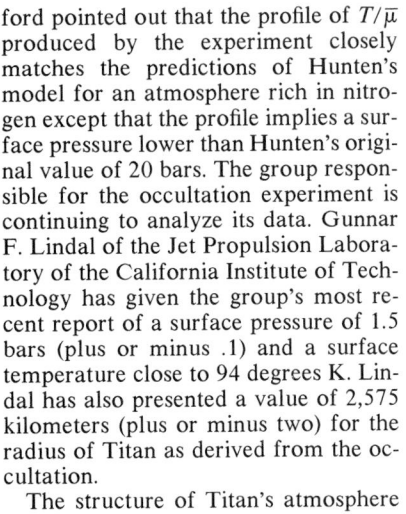

CROSS SECTION OF TITAN'S ATMOSPHERE includes two layers whose presence was discovered by *Voyager 1*. They are a layer transparent to visible light in which ultraviolet radiation is absorbed and below it the layer of high-altitude haze. Below the haze lies the layer of aerosol particles. It is presumed that the particles have been aggregating into larger particles and falling to the surface over the history of the solar system. Methane clouds and methane rainfall are shown above the surface; they are unconfirmed but likely. The curve showing temperature v. pressure (*color*) is based on an experiment in which the atmosphere of Titan intervened between the earth and the radio signals transmitted by *Voyager 1*. According to the data amassed by means of this occultation (along with Voyager data from infrared spectroscopy), the surface temperature on Titan is about 95 degrees Kelvin and the surface pressure is 1,500 millibars (1.5 bars). The average sea-level pressure on the earth is slightly more than one bar.

tan has about 10 times as much gas per unit area of the surface of the body as the atmosphere of the earth.

The infrared spectrometer on *Voyager 1* also recorded emission bands due to several gaseous substances whose presence in Titan's atmosphere had not been previously established. The first of these to be identified was hydrogen cyanide (HCN), a substance that may have been part of the chemical reactions that led to the synthesis of compounds such as adenine on the earth three billion years ago. Adenine is a constituent of DNA; hence it is essential to life on the earth. In subsequent studies Virgil G. Kunde and William Maguire of the Goddard Space Flight Center and their colleagues compared the infrared spectra of Titan with laboratory spectra they had made for that purpose. The comparisons rapidly led to the identification of six further substances in Titan's atmosphere. They include hydrocarbons such as propane (C_3H_8) and nitrogenous com-

pounds such as cyanoacetylene (HC_3N). Darrell F. Strobel of the Naval Research Laboratory has shown that these substances can arise from reactions involving methane and nitrogen that are driven by the bombardment of Titan's atmosphere by ultraviolet photons from the sun and by high-energy electrons trapped in Saturn's magnetic field.

The Surface of Titan

It now seems clear that the early speculation regarding the origin of Titan's aerosol is essentially correct. The molecular fragments and compounds produced by the impact of ultraviolet photons and of high-energy electrons form polymers, or molecular chains. In that way they come to be suspended as solid particles in the atmosphere. By studying the manner in which the aerosol particles reflect sunlight, James B. Pollack and Kathy Rages of the Ames Research Center of NASA could show that the variation in the brightness of Titan with the change in the angle between the sun and Titan and each of the Voyager spacecraft could be explained if the aerosol particles high in the atmosphere have a mean radius of .5 micrometer.

It can be hypothesized that these particles slowly settle out of suspension and that as they sink they collide and aggregate. The aggregates fall faster. Hence the atmosphere steadily loses the carbonaceous and nitrogenous molecules produced by processes high above the surface of Titan. Strobel has estimated that over the age of the solar system a quantity of hydrocarbons that would amount to a layer .1 to .5 kilometer deep has been deposited on the surface of Titan. Along with it has come a deposit of nitrogenous compounds that would amount to a layer some tens of meters deep.

What is the nature of the surface onto which this manna from heaven (as Carl Sagan of Cornell University likes to call the stuff) is falling? Refined analyses of the data from *Voyager 1* place the surface temperature of Titan at 94 degrees K., plus or minus one degree. Moreover, measurements made by the infrared spectrometer on *Voyager 1* suggest that the surface temperature varies by no more than three degrees between the equator and the poles. The reason is the ubiquity of the dense, light-absorbing atmosphere. These values of the surface temperature allow the presence of liquid methane. Indeed, they make it quite possible that Titan is covered by a global liquid ocean of what we on the earth call natural gas.

Methane, therefore, may play the same role on Titan that water plays on the earth. At the surface of Titan the methane is a liquid. In the lower atmosphere it is a gas. Perhaps the lower atmosphere of Titan includes methane clouds, and perhaps the lower atmosphere occasionally becomes saturated with methane in one place or another, so that a methane rain results. At last, it seems, we have found a world besides the earth where a large quantity of some compound is a liquid at the surface. It is too bad for the potential astronaut visitor that the temperature there remains so close to 94 degrees K., or −179 degrees Celsius.

What would it be like to sit in a boat on a methane sea on Titan? The horizontal visibility would be quite good. According to O. Brian Toon of the Ames Research Center, the large aerosol particles dropping out of the atmosphere would be few and far between. The visibility would diminish, of course, in a methane rainstorm. On the other hand, the light would be quite dim. Saturn is nearly 10 times farther than the earth from the sun. That alone diminishes the arrival of sunlight per unit area by a factor of 100. The weakened light would be further attenuated by the aerosol layer and by any methane clouds that happened to be above. It is difficult to estimate how extreme the net attenuation would be. Some of Toon's models predict that the view from the boat would be about as bright as a moonlit night on the earth, even at noon on Titan. Navigation would be hard to manage. The sun and the stars would not be visible. In addition a compass would be useless, because the Voyager spacecraft detected no magnetic field from Titan.

What kind of boat would be appropriate? Probably not a sailboat, because the winds at the surface of Titan are likely to be weak. The reason is the lack of pronounced differences in temperature from place to place. Such differences power the winds on the earth. Could one rely, then, on an outboard motor? Here there is a curious contrast. On the earth the oxidant for an internal combustion engine is freely available in the atmosphere. The fuel, however, is comparatively scarce. On Titan a boat would be afloat on a sea of fuel; it is the oxidant that would be scarce. Perhaps one could get it by drilling for water ice in the interior of Titan and extracting the oxygen from it. Or perhaps outcrops of water ice covered by a layer of hydrocarbons and nitrogenous polymers from the aerosol layer form continents on Titan.

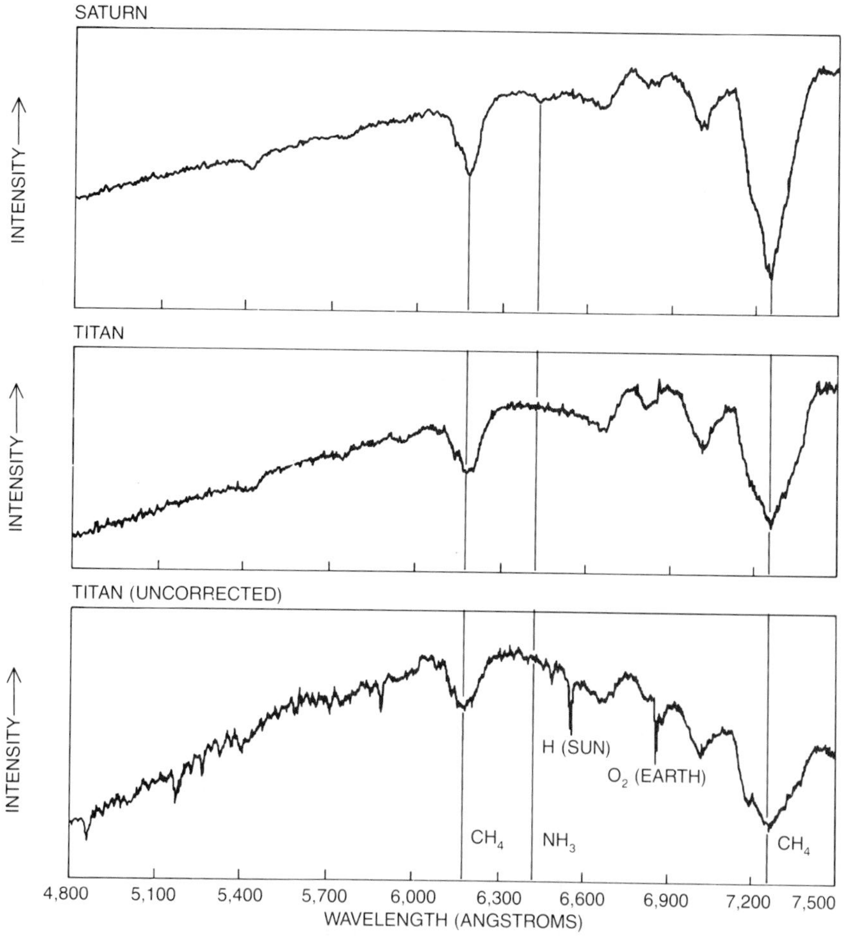

SATURN

TITAN

TITAN (UNCORRECTED)

H (SUN)

O₂ (EARTH)

CH₄   NH₃   CH₄

4,800 5,100 5,400 5,700 6,000 6,300 6,600 6,900 7,200 7,500
WAVELENGTH (ANGSTROMS)

SPECTRA OF SATURN AND TITAN were made by Robert G. Danehy and the author with the aid of the 2.7-meter telescope at the McDonald Observatory in Texas. Blue wavelengths of visible light are at the left of the scale; deep-red wavelengths are at the right. Several absorption bands appear. The one at 6,190 angstrom units and the one near 7,200 angstroms are among those that led to the identification of gaseous methane (CH₄) on Titan. A less prominent band near 6,450 angstroms reveals gaseous ammonia (NH₃) on Saturn but not on Titan. In general the stronger absorption bands in spectra of Titan are shallower and broader than those in spectra of Jupiter and Saturn. The reason is thought to be the scattering of light by Titan's aerosol particles. The top two spectra shown here have been divided by a spectrum of the light reflected from the earth's moon. In this way the spectral lines normally present in sunlight have been reduced to a value of unity and kept from being mistaken for spectral lines of Titan. The bottom spectrum of Titan has not been similarly corrected. It thus shows some spectral lines caused by the absorption of light in the atmosphere of the sun or the atmosphere of the earth.

How the Atmosphere Formed

In the wake of the Voyager missions we must try to understand how the curious atmosphere of Titan evolved. In particular we are faced with a body from which hydrogen escaped, as Jeans could have deduced 65 years ago. In this respect Titan resembles the inner planets of the solar system. Why then has

Titan failed to develop an atmosphere like that of Mars or Venus, an atmosphere rich in carbon dioxide? The reason is that oxygen is unavailable: it is trapped in water ice inside the solid moon. The unique combination of Titan's size and Titan's temperature has allowed the atmosphere of Titan to evolve and yet remain a reducing one.

It is generally accepted that Titan formed in a proto-Saturnian nebula, an isolated part of the cloud of dust and gas that became the solar system. Then too it seems reasonably certain that Titan formed with Saturn, Saturn's rings and Saturn's other moons some 4.5 billion years ago. The density of Titan measured today (1.9 grams per cubic centimeter) indicates that it consists of approximately 52 percent rock and 48 percent ices. The proportions represent a slight enrichment in rock compared with the composition of the solar system overall. The ice, however, would have been crucial for the subsequent evolution of Titan's atmosphere, because the ice would have trapped gases from the proto-Saturnian nebula to a far greater extent than the rock would have.

Twenty years ago Stanley L. Miller of the University of California at San Diego predicted that the icy moons of Saturn should include methane hydrate ($CH_4 \cdot 7H_2O$), that is, methane trapped in water ice. The known presence of methane in Titan's atmosphere supported his idea. The newly discovered presence of several additional gases suggests that they too were trapped as hydrates. In order to predict with assurance which substances really were trapped, one must know the values of temperature and pressure for which a given substance and its hydrate are in equilibrium. (At equilibrium the rate at which molecules or atoms of a substance escape from the hydrate equals the rate at which they are trapped, so that the quantity of the hydrate does not diminish and the hydrate is stable.) It is thought that the proto-Saturnian nebula's temperature did not fall much below 60 degrees K. At that temperature the equilibrium pressure for nitrogen molecules or argon atoms and their respective hydrates is less than 10^{-7} bar.

A pressure of 10^{-7} bar is lower than the pressure that nitrogen or argon is likely to have contributed to the proto-Saturnian nebula; therefore these gases should have been trapped in ices. On the other hand, the equilibrium pressure for neon atoms and their hydrate at 60 degrees K. is nearly 40 bars. Neon has a high cosmic abundance; hence the absence of a detectable amount of it in Titan's atmosphere means two things. First, neon could not be trapped as a hydrate; second, it was not trapped as a gas. That is, the gravitation of Titan as it was forming was not strong enough to trap neon directly from the proto-Saturnian nebula. (The atomic weight of the

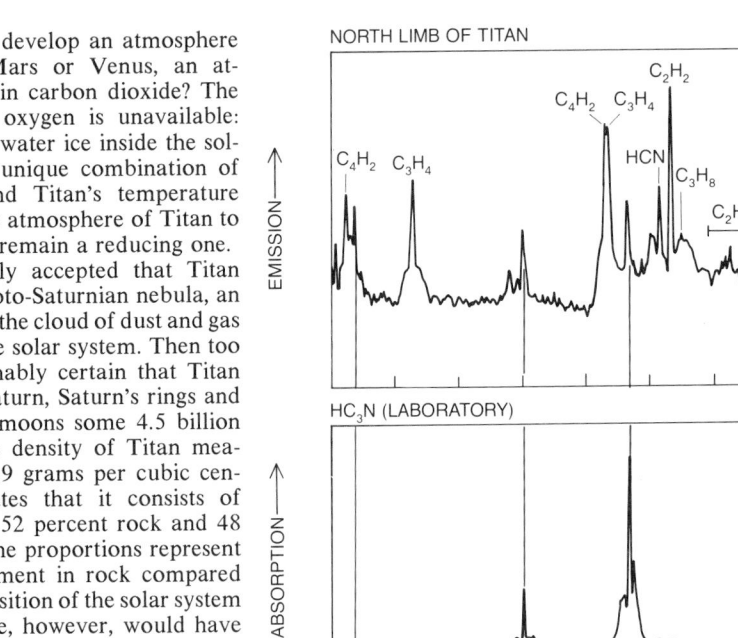

NORTH LIMB OF TITAN

SPECTRA MADE BY VOYAGER 1 in the infrared part of the electromagnetic spectrum allow the identification of several gases on Titan other than methane. In this case the identification of cyanoacetylene (HC_3N) and cyanogen (C_2N_2) is demonstrated by a comparison of absorption spectra made in the laboratory with an emission spectrum of Titan made by the spacecraft. The comparison is valid because the molecules of a given gas absorb and emit radiation at the same set of characteristic wavelengths. Spectral features of several other gases identified by similar comparisons are labeled in the spectrum of Titan. For all three spectra the horizontal scale represents wave number, or waves per centimeter. A wave number of 200 corresponds to a wavelength of 500,000 angstroms; a wave number of 1,400 corresponds to a wavelength of about 71,000 angstroms. The spectroscopy was done by a group led by Rudolf Hanel of the Goddard Space Flight Center of the National Aeronautics and Space Administration.

most abundant isotope of neon is 20, or four more than the upper limit set by Jeans's theory for escape from Titan at its present mass.) The absence of neon tends to confirm that Titan's atmosphere formed after the body itself accreted and that the atmosphere formed from gases trapped as hydrates.

How did the gases escape from the hydrates and get to the surface of Titan? In the first place the release of gravitational potential energy in the form of heat as Titan accreted would have been sufficient to vaporize a fraction of the ices in the body. Later the decay of radioactive nuclei inside Titan would have become the main source of heat inside it. According to models proposed by Mark Lupo and John S. Lewis of the Massachusetts Institute of Technology, the radioactive heating may have been sufficient to create a zone of liquid water deep in the mantle of the body. Gases could escape from the liquid.

Plainly there are ways in which gases once trapped in Titan's ices could escape and form an atmosphere. One sees evidence of such escape on other Saturnian moons. The cracks on Dione rimmed by material brighter than the surrounding terrain are the most conspicuous example. Dione was simply too small to retain an atmosphere. Other Saturnian moons show signs of fresh surfaces. The material that resurfaced the moons may have been driven upward to the surface partly by the internal pressure of gases escaping from hydrates.

The nitrogen in Titan's atmosphere calls for further discussion. In the events I have been describing it is assumed that the nitrogen in Titan's atmosphere today was incorporated into the accreting Titan as a hydrate. This assumption requires in turn that the dominant form of nitrogen in the proto-Saturnian nebula was molecular (N_2), which may not have been the case. To be sure, Ronald G. Prinn of M.I.T. has joined with Lewis in suggesting that N_2 was the stable form of nitrogen in the incipient solar system. Prinn and M. Bruce Fegley, Jr., of M.I.T. point out, however, that the increase of temperature near the incipient Jupiter and Saturn could have allowed ammonia (NH_3) to form. If it did and if Titan trapped it as a hydrate instead of trapping N_2, the subsequent history of Titan must have been substantially different.

In particular, calculations made by Sushil K. Atreya and his colleagues at the University of Michigan show that one must postulate a "warm" epoch early in the history of Titan in which the surface temperature exceeded 150 degrees K. Throughout this epoch ammonia would have escaped from the interior and into the atmosphere, where it would have been broken up by solar ultraviolet photons. In this way the atmosphere would have lost its ammonia and gained the N_2 it has today. A temperature of 150 degrees K. is not out of the question; in principle a greenhouse effect created by hydrogen and ammonia in Titan's early atmosphere could have produced it. Still, the warm epoch is a complication that is avoided if nitrogen was trapped by Titan as a hydrate of N_2.

Perhaps a combination of the two processes was involved, so that less of a constraint need be placed on the early surface temperature.

The argon in Titan's atmosphere also merits discussion. If it makes up 12 percent of the atmosphere, it must have come from trapped hydrates. The only competing possibility is that it came from radioactive decay. More than 99 percent of the argon in the atmosphere of the earth was made in just that way: it arose from the decay of the radioactive isotope potassium 40 into argon 40. Even if the rocky material in Titan had the same proportion of potassium as the rocky material in the earth, however, the current content of argon 40 in Titan's atmosphere would be about 70 parts per million, far from the predicted 12 percent, or 120,000 parts per million. The argon in Titan's atmosphere must indeed be primordial argon trapped as a hydrate from the proto-Saturnian nebula. It must therefore be argon 36 with an admixture of 20 percent argon 38.

The equilibrium pressures of nitrogen and argon with respect to their hydrates are sufficiently similar and sufficiently low to suggest that the proportions in which they become trapped as hydrates should be about the same as the proportions in which they later escape from

those hydrates. A test of the presence of nitrogen instead of ammonia in the proto-Saturnian nebula begins, therefore, with an estimate of the amount of nitrogen that has escaped from the solid body of Titan. The estimate must include the nitrogen content of Titan's atmosphere today and also take account of how much nitrogen has escaped into space and how much has been incorporated into aerosol particles over the history of the body. It follows from Strobel's values for these two modes of depletion that the total amount of nitrogen that has entered Titan's atmosphere is about 1.7 times the atmosphere's current content of nitrogen, or 140 percent of the total content of Titan's atmosphere.

The amount of argon that has escaped from the solid body of Titan is easier to calculate. Argon is inert (it will not form chemical compounds), and it is too heavy to escape into space in appreciable quantity. The amount of argon that has escaped from the solid body is simply the 12 percent of the atmosphere. The ratio of nitrogen to argon is therefore 11.7. The ratio of nitrogen to argon in the cloud of matter that became the solar system was quite close to that: it was 11.

What can Titan tell us about the primitive earth? Opinion is shifting away

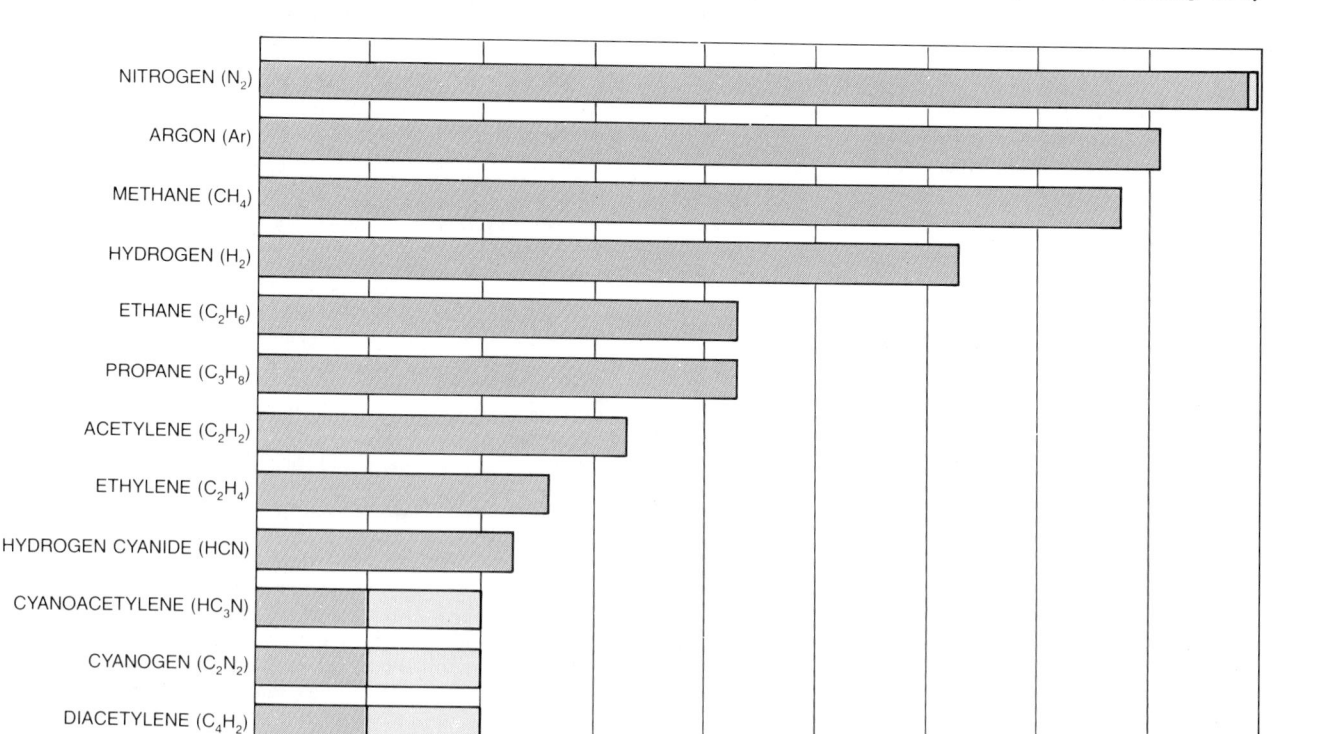

GASES IN TITAN'S ATMOSPHERE are now thought to vary in abundance from molecular nitrogen (set at 82 to 94 percent of the atmosphere, or 820,000 to 940,000 parts per million, as a result of Voyager data) to trace quantities of hydrocarbons such as methylacetylene and nitrogenous substances such as cyanogen. Several further trace constituents of Titan's atmosphere may remain to be discovered. The chart indicates some 12 percent (120,000 parts per million) of the inert gas argon. This 12 percent is required to raise the mean molecular weight of the gases that make up Titan's atmosphere to the value of 28.6 that tentatively emerges from the Voyager data.

from the view that the earth began with a highly reducing atmosphere consisting mostly of methane, ammonia, hydrogen and water vapor. The newer view is that the atmosphere was mildly reducing at first and that it consisted mostly of carbon dioxide, molecular nitrogen and carbon monoxide with no more than 10 percent of hydrogen. It is firmly agreed, however, that no free oxygen was present, because an oxidizing environment makes it extremely difficult for organic molecules to form. In an oxidizing environment the carbon is quickly locked into molecules no more complex than carbon dioxide.

Investigators working in laboratories on the earth have subjected gaseous mixtures of methane, ammonia, hydrogen and water vapor to repeated electric discharges. They have succeeded in producing, among other organic molecules, some of those that are constituents of DNA. What they have not been able to produce is something like DNA itself, or more broadly a molecule that can replicate itself by acting as a template for the assembly of a duplicate from the simpler molecules that are available. The problem seems to be a lack, in Andrew Marvell's phrase, of "world enough, and time." A test tube is simply too small and years may be too short for the advent of a self-replicating molecule to be likely.

Titan as a Laboratory

In this respect Titan offers what amounts to an immense natural laboratory free of experimental bias and the possibility of accidental catalysis or inhibition of chemical reactions by the walls of reaction vessels or the grease in the stopcocks of a vacuum system. Moreover, it is a laboratory where the reaction products over the history of the solar system have been collected into aerosol particles that rest at low temperature at the surface of the body, where we may one day be able to examine them at our leisure. Indeed, we may one day resolve to conduct experiments there ourselves. Heating up a few acres of Titan would produce more of the "primordial soup" than our terrestrial experiments could make in several lifetimes.

For now, NASA is contemplating a mission for the 1990's that would send a probe into Titan's atmosphere while a parent spacecraft was in orbit around Saturn, employing radar to pierce the opacity of Titan's atmosphere and make a map of the surface. We may expect much sooner than that to learn more about Titan from further analysis of the Voyager data, from the powerful new spectrometers now available on the earth and particularly from the instruments that will go into orbit around the earth with the Space Telescope in 1985.

Meanwhile the Voyager spacecraft

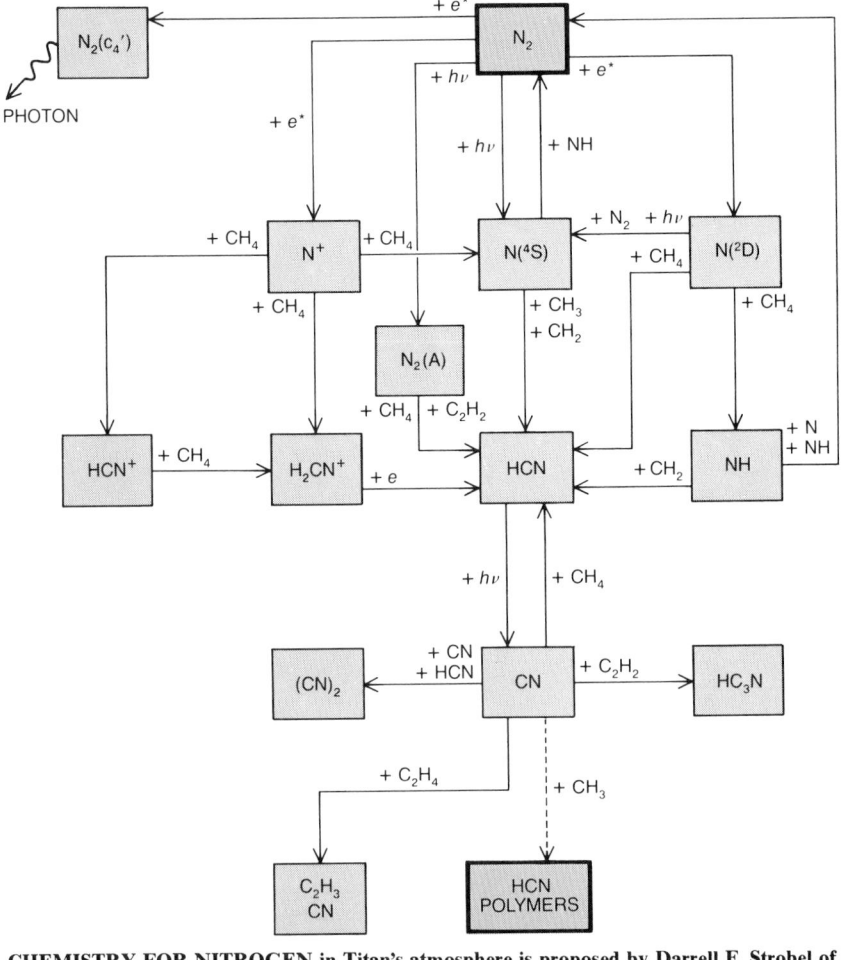

CHEMISTRY FOR NITROGEN in Titan's atmosphere is proposed by Darrell F. Strobel of the Naval Research Laboratory. Here the main reactants and reaction products are shown; hydrogen atoms and molecules, among others, are omitted. The initial reactions involve the breaking of the N–N bond in a nitrogen molecule high in the atmosphere. The energy to break the bond is provided by the impact of solar ultraviolet radiation ($h\nu$) or of a high-energy electron (e^*) trapped near Titan by Saturn's magnetic field. (The latter process supplies Titan's atmosphere with about 10 times as much energy as the former.) The final reactions include the formation of the reddish aerosol particles observed in Titan's atmosphere. The final steps are not fully known because the chemical composition of the particles themselves is not yet established. It is, however, known that hydrogen cyanide (HCN) can form a reddish-brown polymer. In the illustration ^{2}D and ^{4}S signify excited atomic states, A and c_4' signify excited molecular states and e signifies an electron freed in the atmosphere by solar ultraviolet radiation.

continue their missions of exploration. *Voyager 1* is on its way out of the solar system. If its transmitters continue to function we can expect it to tell us the position of the boundary between the solar wind and the tenuous gas of the interstellar medium. The news should come within the next 10 years. *Voyager 2* is on its way to Uranus, which it will reach in January, 1986, and Neptune, which it will reach in August, 1989. The mechanism that points the cameras on the spacecraft suffered some damage when the spacecraft passed through Saturn's rings. Engineers at the Jet Propulsion Laboratory are confident, however, that *Voyager 2* will transmit valuable data from Uranus and Neptune if there are no further equipment failures.

Still, the Voyagers were not originally intended to travel to planets more distant than Saturn, and their instruments are far from optimum for that purpose.

The change in plans was made in the mid-1970's when it became apparent that no specific missions to Uranus and Neptune would be undertaken for at least a decade. In the wake of the success of the Voyager program it is ironic that we find ourselves with no follow-on missions other than one that may be canceled: the Galileo project, which would send a probe into Jupiter's atmosphere to make measurements of such things as composition, pressure and temperature. We have already lost our chance to send a mission to Halley's comet. In effect we are retreating to the early 17th century, a time when the most distant known planet was Saturn and it was not understood that comets traverse the solar system. This is no time for such a retreat. There is much we need to learn about the worlds around us before we can fully understand the history of the solar system and that of the earth itself.

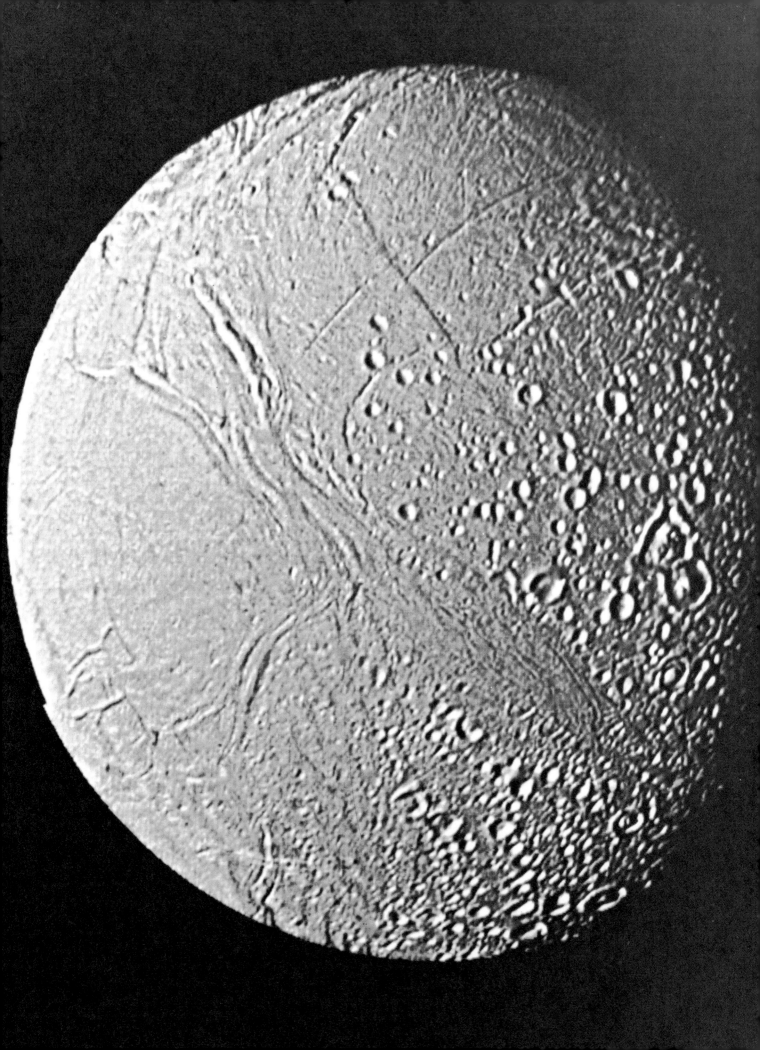

The Moons of Saturn

9

by Laurence Soderblom and Torrence V. Johnson
January, 1982

*The 17 icy bodies that orbit the planet display a
surprising range of geological evolution. Many of them
show craters more than four billion years old, but one of
them has terrain so new that no craters are seen*

Before the spacecraft *Voyager 1* neared Jupiter in March, 1979, only five bodies in the solar system other than the earth had been observed in sufficient detail for their history to be surmised. In essence all of them (Mercury, the moon and Mars and its moons Phobos and Deimos) consist of rocky material. The encounter with Jupiter and its moons doubled the list. Moreover, it marked the first appearance on the list of planet-size moons composed mostly of ices.

The encounters with the Saturn system doubled the list again. In November, 1980, *Voyager 1* flew past Titan, the largest moon of Saturn, at a distance of only 7,000 kilometers. It passed the smaller moons Mimas, Dione and Rhea at greater distances but nonetheless transmitted high-resolution images of each back to the earth. The trajectory of *Voyager 2* had already been devised to bring the spacecraft closer than *Voyager 1* had come to the moons Iapetus, Hyperion and Phoebe; in addition it would come very close to Tethys and Enceladus. In the months before the arrival of *Voyager 2* near Saturn last August the sequence of observations planned for the spacecraft was altered to provide for observations of several newly discovered moons, three of which were found by *Voyager 1*. In spite of a temporary jam in the mechanism that points the

cameras on *Voyager 2*, the mission was successful. In a few short months, therefore, the moons of Saturn have been transformed. Before November, 1980, the ones that were known were no more than dots of light in a telescope. Now they form an array amounting to 17 new worlds.

General Properties

Several generalizations can be made about the moons of Saturn. In the first place only one of them has any appreciable atmosphere. It is Titan, whose atmosphere is opaque to visible light. Since no one has seen the surface of Titan, we have little direct information about its geological evolution. Second, it can be calculated that all but the three outermost moons of Saturn should certainly be in synchronous rotation: they should keep the same face turned toward the planet, just as the moon keeps the same face turned toward the earth. In each such case a planet's gravitation raises a tidal bulge on a moon. Then the gravitational attraction between the bulge and the planet acts as a torque that slows the moon's rotation until the rotation is synchronous. The Voyager images suggest that all but one or possibly two of the moons of Saturn really do rotate synchronously. The definite exception is Phoebe, the outermost

moon, which is too small and too far from the planet to lose its spin to tidal forces. It rotates once every nine hours, whereas its orbital period is 13,211 hours, or 1.5 years. The possible exception is Hyperion, the third-outermost moon. The images made of Hyperion by *Voyager 2* cover only a short arc of the orbit of the moon and leave it uncertain whether Hyperion rotates synchronously or not.

Finally, all but two of the moons of Saturn form a regular system of satellites. That is, their orbits are nearly circular and lie in the equatorial plane of the planet. The two exceptions are Iapetus, the second-outermost moon, whose orbit is inclined 14.7 degrees with respect to the equatorial plane, and Phoebe, the outermost, whose orbit is inclined 150 degrees. (In addition the orbital motion of Phoebe around Saturn is in a direction opposite to that of all the other moons.) Three regular systems of satellites are known in the solar system; they consist of the inner moons of Jupiter, the inner moons of Saturn and the five known moons of Uranus. It is probably no coincidence that those planets all have rings. Rings and a regular system of satellites may form naturally as a by-product of the accretion of a giant planet. In any case each regular system of satellites is thought to have formed from the gas, ice and dust around each incipient giant planet, much as (on a larger scale) the planets formed around the sun.

Observations made with telescopes before the Voyager spacecraft arrived showed that among the moons of Saturn only Titan is as large as the four moons of Jupiter discovered by Galileo; the others are smaller than the earth's moon. The Voyager images show that the moons span a range of sizes from the size of asteroids to that of Mercury. The masses of Saturn's moons remain difficult to specify. An analysis of the mutual gravitational effects the moons have on one another has yielded values for some of the masses, and the tracking of gravitational perturbations in the trajectory of the Voyager spacecraft as they

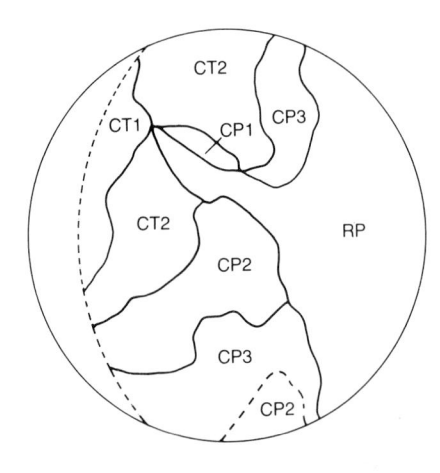

SATURNIAN MOON ENCELADUS was photographed last August 25 by the spacecraft *Voyager 2*. It is some 500 kilometers in diameter and is seen here at a range of 119,000 kilometers in an image that has been processed by computer so that the surface topography stands out. Six different terrains are distinguished. The heavily cratered terrains called *CT*1 and *CT*2 are the oldest; many of their craters are thought to represent the bombardment of Enceladus and other Saturnian moons by debris in orbit around Saturn. The debris was left over from the accretion of the planet, its moons and its rings. The cratered plains *CP*1, *CP*2 and *CP*3 are intermediate in age. The ridged plain *RP* is the youngest; on one hypothesis it lacks visible craters because it consists of upwelling fresh material.

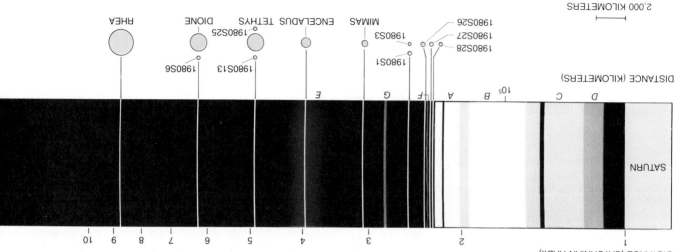

KNOWN MOONS OF SATURN are 17 in number. The top panel shows their orbits. All but two of them lie in the equatorial plane of the planet, which is also the plane of the planet's rings. The exceptions are Iapetus (with an orbit inclined by 14.7 degrees) and Phoebe (with one inclined by 150 degrees). Phoebe's rotation about Saturn is clockwise; all the other moons go counterclockwise. The top panel also shows the trajectory followed by *Voyager 1* in November, 1980, and the one followed by *Voyager 2* last August. The middle panel shows the orbits of the moons on a logarithmic scale; the numbers on the scale are distances from the center of Saturn. Each of the

---

of the values of the densities are consistent with a composition of rock and ice that is similar in all Saturn's moons except for more or less random variations in the exact proportions. On the other hand, the densities of Saturn's moons in general are less than those of Jupiter's. This suggests a greater proportion of ice. The relative lack of rocky material is explained by models of Saturn's history developed by James B. Pollack and his colleagues at the Ames Research Center of the National Aeronautics and Space Administration. The models suggest that rocky material near Saturn was swept into the incipient planet as it contracted before its moons began to form some 4.5 billion years ago.

In any event the surfaces of the moons of Saturn suggest the presence of ice.

In one sense the calculated densities are curious. Among the planets of the solar system one finds a trend toward greater density with decreasing distance from the sun, and among the moons of Jupiter one finds a trend toward greater density with decreasing distance from the planet. Such trends are attributed to the influence of the heat from the central body on the temperature of the gas and dust that surrounded it at the time its satellites formed. In the case of Jupiter it appears that ice was unstable in any large quantities at radial distances from the planet less than the distance of the orbit now followed by Ganymede. In spite of the uncertainties in the measurements no similar trend of density is evident among the moons of Saturn. Instead the current determinations

passed among the moons yields new values and refinements of some of the earlier ones. It seems clear from the various measurements that the moons of Saturn all have densities of less than two grams per cubic centimeter. In fact, several have densities of less than 1.5 grams per cubic centimeter. Such values suggest that the moons are composed mostly of ice. For most of Saturn's moons a composition of 30 to 40 percent rock and 60 to 70 percent ice by weight would match the calculated density. Only Titan is large enough for its gravitational self-compression to affect its density appreciably. When this self-compression is taken into account, the density estimated for Titan—1.9 grams per cubic centimeter—becomes compatible with a mixture of half rock and half ice.

Spectra of the solar radiation reflected by the moons in the near infrared show its absorption at wavelengths characteristic of water frost. Moreover, measurements made by the Voyager spacecraft show that most of the moons reflect between 60 and 90 percent of the radiation that hits them. With an albedo, or reflectance, of nearly 100 percent, Enceladus is the most reflective body in the solar system. If it were at the same distance from the sun as the earth's moon, it would be about five times as bright. Water ice is highly reflective.

The ubiquity of ice in the outer solar system was not entirely unexpected. For one thing, the vapor pressure of water ice (that is, the tendency of the ice to sublimate and lose vapor to space) depends strongly on temperature. Hence at distances from the sun less than the distance of the asteroid belt between the orbits of Mars and Jupiter an unprotected mass of ice will evaporate in a time quite short compared with the age of the solar system. At greater distances a mass of ice will be stable for billions of years. In addition most models of how the solar system formed predict that water should be a major constituent of a body that accreted at low temperatures.

Specifically, if a gas whose composition is much like that of the sun is cooled under the conditions of temperature and pressure thought to have prevailed in the early solar system, some of the oxygen in the gas will combine with silicon to form silicate rock at relatively high temperatures. When the elemental silicon is exhausted, however, a substantial amount of oxygen will remain. As the temperature continues to decrease it will combine with hydrogen, the element most abundant in the gas. Thus water will form. It emerges from such models that a moon condensing at low temperatures should have much the same proportions of rock and ice as those inferred for Jupiter's moons and Saturn's moons from the estimates of their densities.

One might suppose an icy moon would be no more interesting geologically than a cratered ice cube. Ice, however, has a melting point far lower than that of rock, so that relatively little is needed to melt the interior of an icy moon in the outer solar system. On this basis it was suspected even before the Voyager missions that such moons

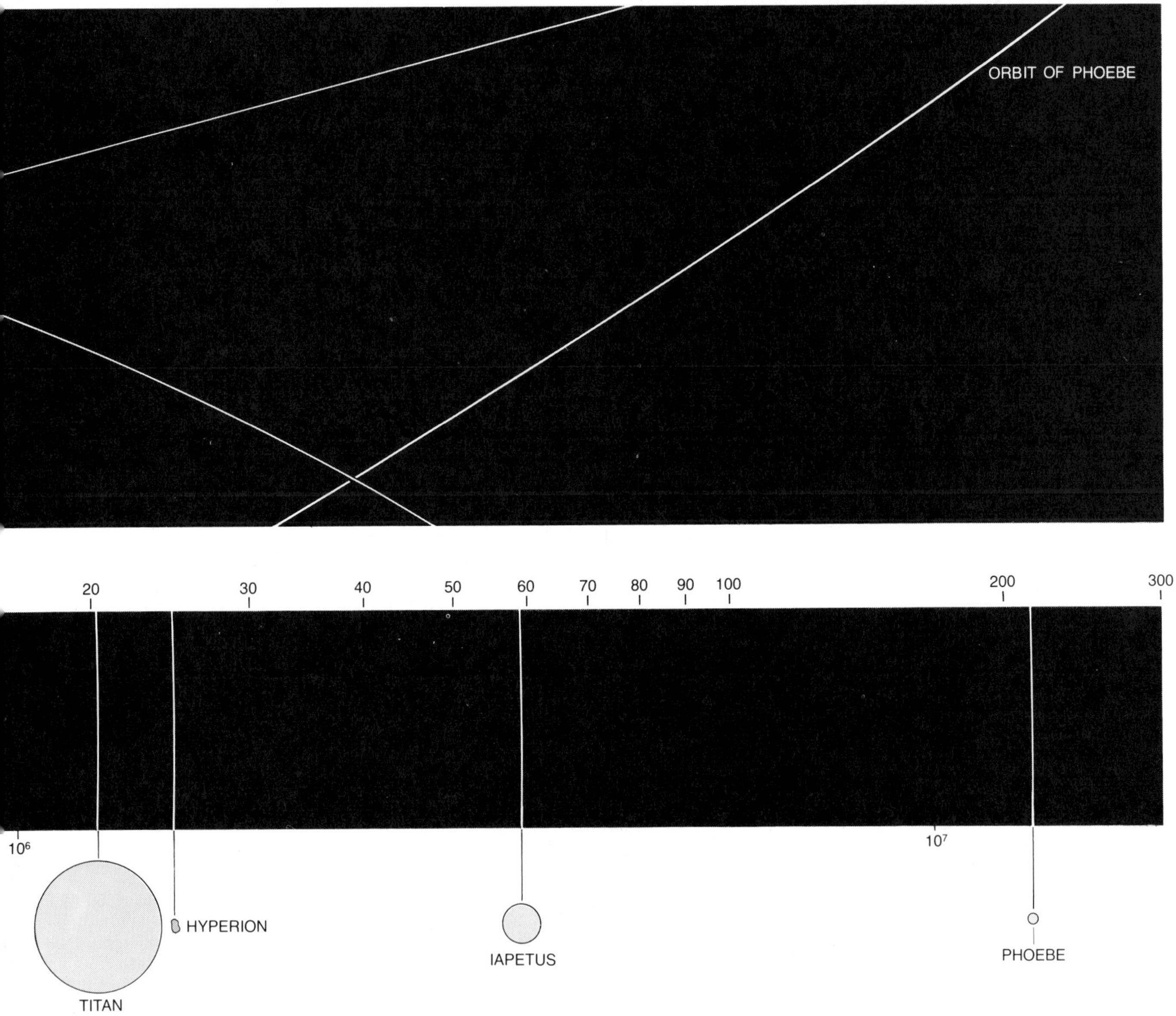

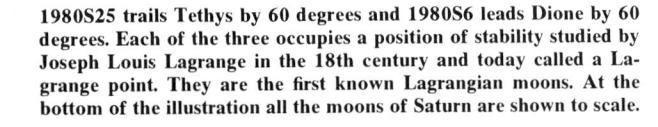

ORBIT OF PHOEBE

| 20 | 30 | 40 | 50 | 60 | 70 | 80 | 90 | 100 | 200 | 300 |

10^6 10^7

TITAN   HYPERION   IAPETUS   PHOEBE

moons discovered recently turns out to have a dynamical vagary. The moon 1980S28 lies near the outer edge of the A ring; 1980S27 and 1980S26 bracket the F ring, and 1980S1 and 1980S3 have orbits that differ by less than the sum of the diameters of the moons (hence they are "co-orbital"). Finally, 1980S13 leads Tethys by 60 degrees, 1980S25 trails Tethys by 60 degrees and 1980S6 leads Dione by 60 degrees. Each of the three occupies a position of stability studied by Joseph Louis Lagrange in the 18th century and today called a Lagrange point. They are the first known Lagrangian moons. At the bottom of the illustration all the moons of Saturn are shown to scale.

might show signs of geological activity. It was also proposed that the moons of the outer solar system might incorporate substances such as ammonia hydrates and the compounds of methane and water known as clathrates. The interior of a moon incorporating such material would melt even more readily than a moon consisting of rock and water ice alone. The results of the Voyager missions surpass the speculations. Hyperion, Mimas and Enceladus, for example, are much the same size, but they display a range of geological evolution far broader than one would have thought was likely.

The New Satellites

The moons of Saturn discovered in the past decade have dynamics that are unusual in one way or another. Consider 1980S28, the innermost moon in the set and also the innermost known moon of Saturn. It was discovered by *Voyager 1* just beyond the outer edge of Saturn's *A* ring. 1980S28 is an elongated body whose diameter is about 40 kilometers. One hypothesis suggests that its gravitational field does much to sculpture the sharp outer edge of the ring. Somewhat farther out from Saturn are the pair of small moons 1980S27 and 1980S26. They too were discovered by *Voyager 1*, although it can now be recognized that at least one of them may have affected the counts of charged particles made by the *Pioneer 11* spacecraft near Saturn in September, 1979. Between their orbits lies the multistrand *F* ring. The gravitational fields of the moons may well confine the ring. For that reason they are called the shepherd moons.

About 10,000 kilometers beyond the *F* ring, or roughly halfway between the *F* ring and the *G* ring, are the "co-orbital" moons 1980S1 and 1980S3. The French astronomer Audouin Dollfus photographed one of them at the Pic du

Midi Observatory in 1966. Today it is difficult to say which one it was. In 1978 John W. Fountain and Stephen M. Larson of the University of Arizona determined that there were two moons. Then in 1979 *Pioneer 11* made an image of one of them. The Voyager spacecraft made images of both. The mean orbital radii of 1980S1 and 1980S3 differ by less than the sum of their diameters. Hence their orbital velocities are similar but not quite identical: the inner moon slowly overtakes the outer one. As they approach each other the gravitational attraction between them alters their angular momentum. The inner moon gains momentum; it moves into a larger orbit, where its orbital speed is reduced. The outer moon loses momentum; it moves into a smaller orbit, where its orbital speed is increased. In short, the two moons change places; the inner moon becomes the outer one and begins to fall behind. About once every four years the celestial dance is repeated and the two change places again.

Three other new moons of Saturn were found by earth-based telescopes in 1979 and 1980, a time when Saturn's rings were edge on as they are seen from the earth. This orientation greatly reduces the effect of scattered light from the rings when Saturn is viewed in a telescope and allows the detection of faint bodies close to the planet. Two of the three new moons share their orbit with Tethys. One of them maintains a position about 60 degrees ahead of Tethys; it was discovered by a group led by Bradford A. Smith of the University of Arizona. The other stays about 60 degrees behind Tethys; it was discovered with the aid of a prototype of a planetary camera system designed for the telescope that the U.S. plans to put in orbit around the earth. The third moon stays about 60 degrees ahead of Dione; it was discovered by two French astronomers: P. Lacques of the Pic du Midi Observa-

tory and J. Lecacheux of the Paris Observatory.

Each of the positions occupied by these moons is a point of dynamical stability of the kind first studied by the French mathematician Joseph Louis Lagrange and today called a Lagrange point. In 1772 Lagrange noted that in a system consisting of one body in orbit around another (say the moon around the earth) there are five positions at which a third body can lie undisturbed. Three of the positions are unstable: a body at any one of them is readily driven from it by the influence of gravitational forces other than those exerted by the two bodies that set up the system. The three positions lie (1) inside the orbit of the satellite, (2) outside the orbit and (3) at the point in the orbit opposite the satellite itself. The remaining two positions—the ones on the orbit 60 degrees ahead of the satellite and 60 degrees behind—are quite stable: a body that occupies either one of them will merely drift back and forth along its orbit under the influence of perturbing forces. It has long been known that the groups of asteroids called the Trojans occupy the two Lagrange points 60 degrees from Jupiter along the orbit of the planet around the sun. The three bodies at Lagrange points near Saturn are the first known Lagrangian moons.

In contrast to their dynamical vagaries the newly discovered moons have much the same appearance. Each is rather small, and almost all of them are quite irregular in shape. The irregularity tells much. It suggests that each moon arose from the fragmentation of a larger body. And since a small, cold, icy moon is strong enough to keep itself from being pulled into a sphere by its gravitational self-attraction, the irregular shapes suggest that these moons have not been heated significantly since the time they took their current form.

Iapetus

Iapetus and Rhea are the second-largest of Saturn's moons; they have a nearly identical diameter of about 1,500 kilometers. They move in quite different orbits. The orbit of Iapetus lies about 60 Saturnian radii (R_s) from the center of Saturn, that of Rhea about 9 R_s. They bracket the orbits of two other moons of Saturn, namely Titan and Hyperion. As we have noted, the orbit of Iapetus is inclined.

Iapetus has a density of about 1.1. This means its density is almost as low as that of pure water ice. Since it keeps the same face turned toward Saturn, its trailing hemisphere (the side facing backward with respect to its orbital motion) is always the same part of the surface. The trailing hemisphere is bright. In the visible region of the electromagnetic spectrum its albedo is almost 50 percent. In contrast the leading hemi-

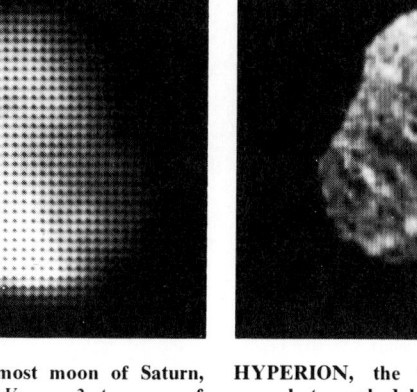

PHOEBE, the outermost moon of Saturn, was photographed by *Voyager 2* at a range of 2.2 million kilometers. Its curious orbit and its low reflectivity (about 5 percent) suggest that it formed elsewhere in the solar system and then was captured by Saturn's gravity. Phoebe is about 200 kilometers in diameter.

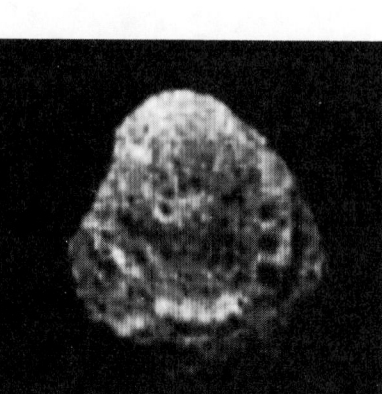

HYPERION, the third-outermost moon, was photographed by *Voyager 2* at a range of 500,000 kilometers. It is an irregular body about 400 kilometers long and 220 kilometers wide. It is four to six times brighter than Phoebe. The concavities along the margin of this image of Hyperion are probably craters.

sphere is extremely dark: its albedo is only 3 to 5 percent. This difference was noted in the 17th century by Jean Dominique Cassini, the discoverer of Iapetus, who found he could see the body on one side of Saturn but not the other. Among the few materials whose albedo is only a few percent are lampblack and the primitive meteorites called carbonaceous chondrites.

The peculiar pattern of darkness and brightness on Iapetus has suggested that dark material falling from space coats what amounts to the prow of the moon. In 1974 it was further suggested by Steven Soter of Cornell University that the source of the material is Phoebe. Phoebe had long been taken to be a moon of low albedo; observations made with earth-based telescopes show that its color is similar to that of many dark asteroids. The idea, then, is that dark matter is kicked off Phoebe by the impact of micrometeoroids. Then the particles of the matter come under the influence of what is called the Poynting-Robertson effect. Specifically, electromagnetic radiation leaves each particle because the particle reflects some of the radiation incident on it and because the particle absorbs some radiation and reemits it later. In either case the radiation leaving the particle in the direction of the particle's orbital motion around Saturn undergoes a Doppler shift that gives it a higher frequency (and thus more energy and momentum) than that of the radiation leaving in the opposite direction. The net result is that the particle loses orbital angular momentum and slowly spirals inward. As it falls (the argument concludes) it is swept up by Iapetus.

The images of Phoebe made by *Voyager 2* show some detail of Phoebe's sur-

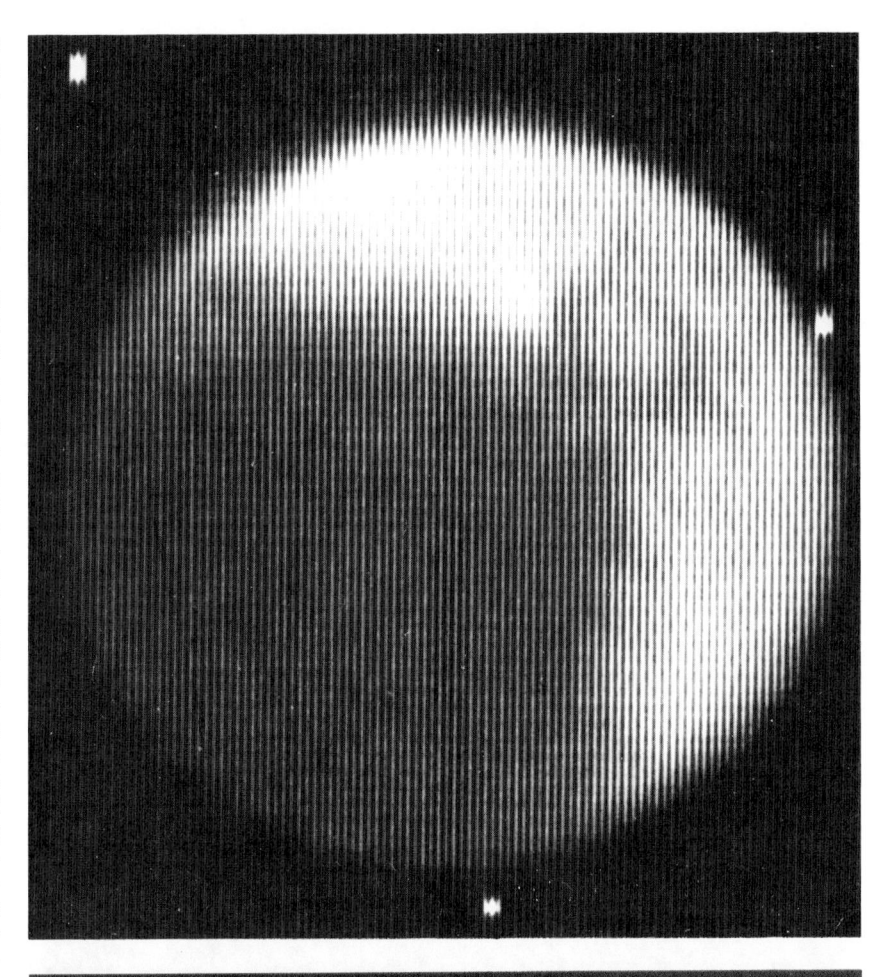

IAPETUS was photographed by *Voyager 1* at a range of 3.2 million kilometers and nine months later by *Voyager 2* at a range of 1.1 million kilometers. The image made by *Voyager 1* (*top*) shows the side of the moon that always faces Saturn. (Iapetus and most other Saturnian moons keep the same face turned toward Saturn, just as the earth's moon keeps the same face turned toward the earth.) Iapetus is moving toward the left; the leading hemisphere of Iapetus (the hemisphere facing in the direction of the orbital motion of the moon) is dark. A ring of dark material extends, however, into the bright trailing hemisphere. The image made by *Voyager 2* (*bottom*) shows the side of the moon that always faces away from Saturn. The north pole of the moon coincides roughly with the large crater astride the border between day and night on Iapetus (the terminator), which crosses the top of the image. The dark leading hemisphere is at the right. An equatorial band of darkness extends into the trailing hemisphere. The sharp and complex boundary between bright and dark regions on Iapetus militates against the proposal that dark matter fell from space onto the leading part of the moon. It suggests instead that the dark matter is an extrusion from within Iapetus.

face. Phoebe emerges as being roughly spherical and having an albedo of only about 5 percent. In both of these attributes (and in its inclined, retrograde orbit) it proves to be quite different from the co-orbitals, the ring shepherds and the Lagrangian moons of Saturn. Phoebe may indeed be a captured dark asteroid, unchanged since it accreted early in the history of the solar system. Perhaps it is a primordial body that was ejected from the inner solar system by the gravitational field of the growing planet Jupiter and took up an orbit around Saturn.

In themselves the Voyager images of Phoebe cannot make it plain whether Phoebe dusted the dark leading hemisphere of Iapetus; the idea is still being debated. One test of the idea involves Hyperion, the next moon inward from Iapetus. The test rests on the assumption that Hyperion should have been dusted by material kicked off Phoebe that Iapetus failed to sweep up. The images made by *Voyager 2* show Hyperion to be a remarkable little moon. It is nearly the size of Mimas, a spherical moon with a radius of about 200 kilometers, and yet it is markedly elongated: its short axes are only about three-fifths the length of its longest axis. Hyperion is one of the largest bodies in the solar system that has an irregular shape. (The asteroid Hector is one of the very few other such objects.) An analysis of the Voyager images indicates that Hyperion is somewhat darker than most of the moons of Saturn: its albedo is 20 to 30 percent. This finding is consistent with a dusting by matter from Phoebe. Hyperion appears, however, to lack a dark leading hemisphere. The problem with searching for one is that it is not yet established whether Hyperion rotates synchronously.

An answer to the question of whether Phoebe dusted Iapetus therefore awaits in part the determination of the rotation rate of Hyperion from observations made with ground-based telescopes. Meanwhile the findings of Dale P. Cruikshank and his associates at the University of Hawaii at Manoa may have got the idea into trouble on other grounds. Cruikshank's group has found from telescopic observations that the color of Phoebe and the color of the dark leading hemisphere of Iapetus are different. The dark regions on Iapetus are much redder than Phoebe throughout the visible and the near-infrared region of the spectrum. This makes it difficult to favor a scheme in which material from Phoebe simply coats the dark part of Iapetus without having undergone some kind of change.

The images of Iapetus made by the Voyager spacecraft also tend to subvert the hypothesis that it was dusted by matter from Phoebe. *Voyager 1*'s best image of Iapetus was made at a range of 2.5 million kilometers. The image showed only those features on Iapetus that are more than 50 kilometers across. Even so, a clue to the history of the body emerged. A ring of dark material about 100 kilometers in diameter was found to straddle the border between the hemispheres. It strongly resembles the rings in craters along the margin of large volcanic flood plains on the moon and Mars. Such rings formed there when dark volcanic melts flowed into impact craters and filled each one around its central peak. Perhaps the dark ring on Iapetus formed by a similar process when fluid extruded from the interior of the body darkened half of it. Clearly it is unlikely that a feature with such a peculiar geometry as that of a dark ring could have been formed by matter falling from space.

Voyager 2 made a series of images of Iapetus at a resolution three times greater than that of the images made by *Voyager 1*. The single best image is of the north polar region, mostly in the bright trailing hemisphere. It shows that the trailing hemisphere is heavily cratered. Many of the craters have a dark floor. Such floors resemble the dark floors of craters on the highlands of the earth's moon, which are thought to have formed when volcanic material flowed out over them. Taken together, the dark-floored craters in the trailing hemisphere and the sharp definition and complexity of the boundary between dark and bright terrain imply a history of eruptions from the interior of Iapetus. That is not to say the erupting material resembles any ordinary lava. One can speculate that it is a fluidized slurry of a mixture including ammonia, ice and something dark. The low albedo of the material suggests that whatever its origin, the dark stuff is rich in organic substances such as are found in primitive meteorites.

Rhea

The distant views of Rhea transmitted to the earth by *Voyager 1* as the spacecraft approached the moon showed a bright, bland leading hemisphere marked only by what appears to be a large and relatively recent impact crater. The trailing hemisphere is different. It shows a complex pattern of bright swaths on a background darker than the leading hemisphere. The swaths are thought to result from internal activity. Perhaps the bright material was extruded along lines of fracture at the surface. In any event the bright swaths fail to follow the kind of pattern that cratering would lay down on the surface. The zone containing the swaths is confined within a small circle at the center of the trailing hemisphere. The boundary be-

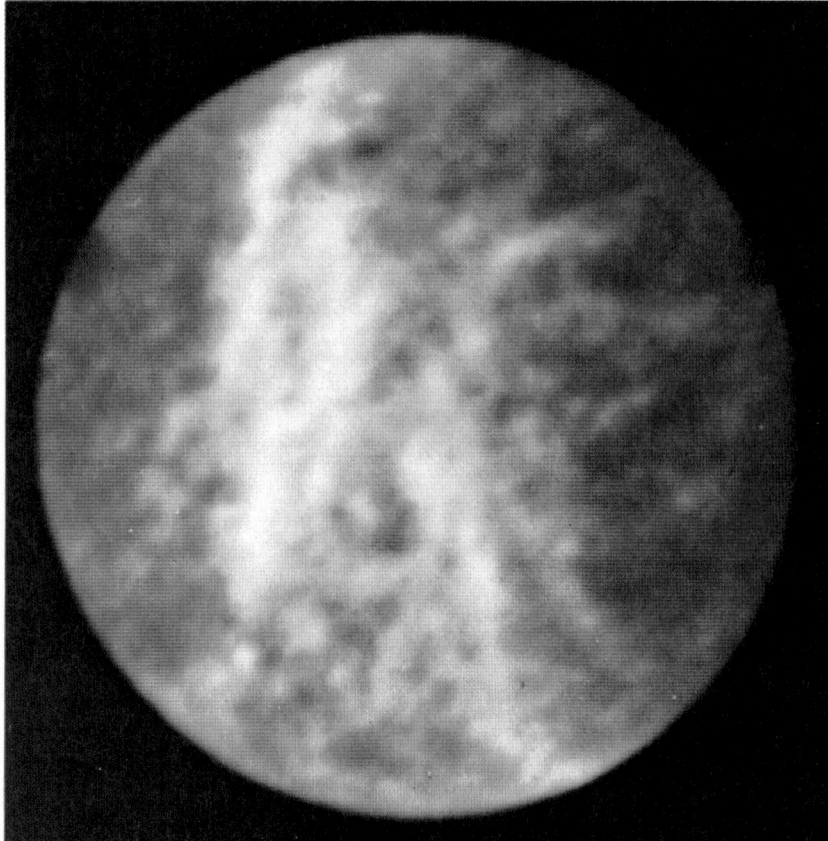

RHEA was photographed by *Voyager 1* **at a range of 1.7 million kilometers. Here the image of the moon is shown in false color. The trailing hemisphere (seen here) is marked by bright wisps that do not seem to follow a pattern laid down by craters. The leading hemisphere is bland.**

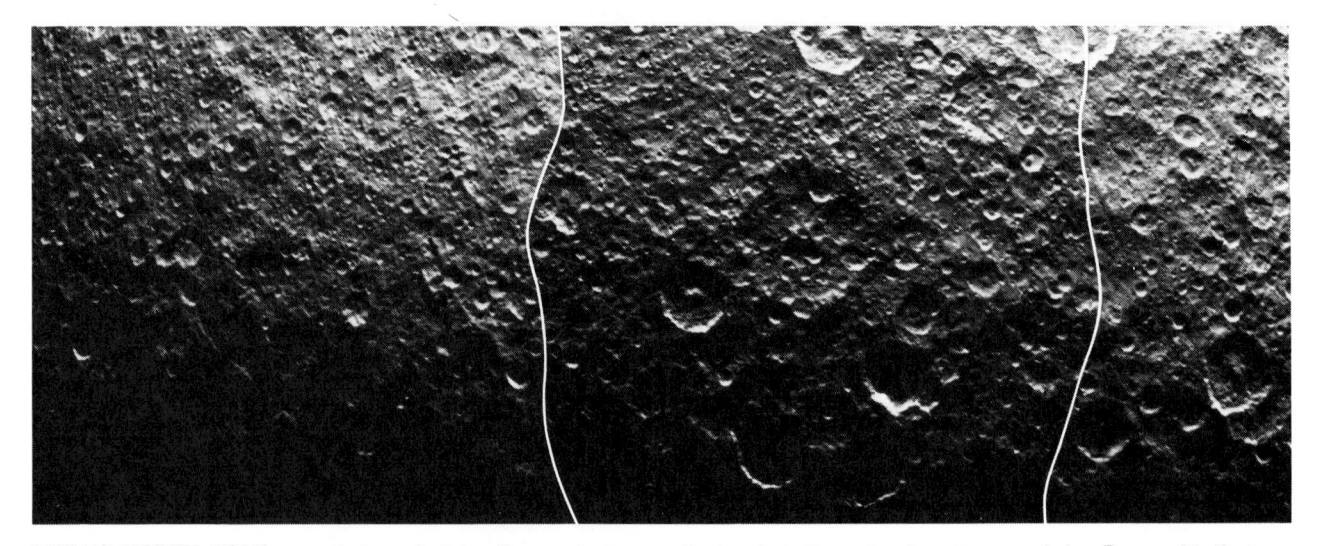

RHEA'S NORTH POLE was photographed by *Voyager 1* **at a range of 80,000 kilometers. The pole itself is at the middle of the arc the terminator describes across this mosaic. The terrain to the west of the pole (toward the right in this image) is marked by large craters; in** the terrain to the east such craters are missing. Presumably the large craters were made early in the history of the moon; the small craters both east and west were made later, after geological activity had given the terrain to the east of the pole a fresh, uncratered surface.

tween this zone and the bland leading hemisphere is diffuse.

The high-resolution views of Rhea made by *Voyager 1* show the equatorial region of the leading hemisphere somewhat to the east of the putative large, fresh impact crater. The best views show the north pole of the body, which *Voyager 1* flew over at a distance of only 59,000 kilometers. In each view Rhea is found to have a densely cratered surface much like the cratered highlands on the moon and Mercury. The principal difference is that the large, fresh craters on the moon and Mercury are surrounded by blankets of ejecta; the craters on Rhea are not. Presumably Rhea's weaker gravity is responsible. If Copernicus, a large crater on the earth's moon, had formed on Rhea, its ejecta would have been spread over most of the body's surface. On the earth's moon most of the ejecta lie no farther away than a few times the radius of the crater.

The nature and the origin of the bodies that cratered Rhea can be surmised from a mosaic of images showing the body's north polar region. At first the entire region appears to be uniformly pitted with craters. A closer examination shows that the western two-thirds of the polar mosaic is marked by a collection of craters with diameters ranging from 30 to 100 kilometers and in addition a dense population of smaller craters. The smallest ones visible are a few kilometers across. If other craters are smaller still, they are below the limit of resolution. The eastern third of the mosaic is also marked by small craters, but the larger craters are missing. A few subtle depressions in the surface are visible. They suggest that the larger craters in the east were filled or buried.

The difference between the two terrains implies a period of cratering in which projectiles with a wide range of energy formed craters with a wide range of sizes. At some point the remaining projectiles whose impacts could form large craters (craters greater than 50 kilometers in diameter) were swept up. The bombardment by the objects that formed the smaller craters continued. Meanwhile part of Rhea's surface was regenerated, perhaps by fluid extrusion from the interior or perhaps by the flow of a slurry of material driven to the surface by the pressure of gases accumulating within the moon. The resurfacing included the eastern third of the polar terrain. Estimates of the rate at which Rhea is being cratered by comets and asteroids today indicate that few of the visible craters (both large and small) in the images are recent. Hence most of the cratering must have come early in the history of the solar system, and it must have been intense. The projectiles could have been remnants of the accretion of gas and dust in orbit around the sun. They could have been remnants of the accretion of gas and dust in orbit around Saturn. They could also have come in pulses as the debris from bodies that collided in the early Saturnian system.

Dione

In sum, the appearance of Rhea suggests that at least two populations of projectiles marked the surface. The appearance of Dione, the next moon inward from Rhea, allows more detailed deductions. Dione travels in orbit around Saturn at a distance from the planet's center of 6.2 R_s. Its diameter is 1,100 kilometers. With a density of 1.5 grams per cubic centimeter it is the densest of Saturn's moons except Titan. The distant views of Dione made by *Voyager 1* showed a striking asymmetry between the leading and the trailing hemispheres. In the trailing hemisphere a network of intersecting bright streaks on a dark background was visible. It looked much like the trailing hemisphere of Rhea. The leading hemisphere was uniformly bright. Again the region of complex variations in albedo was confined within a small circle at the center of the trailing hemisphere, and again the boundary of this region with the bland leading hemisphere was diffuse.

Closer views of the trailing hemisphere made by *Voyager 1* showed that craters between 50 and 100 kilometers in diameter were crossed by the streaks. Therefore the streaks must have formed well after the torrential bombardment the craters imply. All things considered, the large craters on Dione and on parts of Rhea (including the western two-thirds of the polar mosaic) resemble the highland craters found on the earth's moon, Mars and Mercury. We shall refer to the projectiles that made such craters in the Saturnian system as Population I. The most likely hypothesis is that they resulted from the agglomeration of matter left in orbit around the sun after the solar system formed.

The closer views of Dione also showed plains where the density of cratering is far lower than it is on the "highlands." On the plains, however, the proportion of small craters to large ones is far higher. The proportion is that of the eastern third of the polar mosaic of Rhea. On both the plains of Dione and the terrain east of the north pole on Rhea the distribution of sizes among the craters resembles that of the craters surrounding the principal craters on the moon. It also resembles that of the craters surrounding the places on the earth where nuclear devices have been tested. In general the prevalence of small craters suggests the pocking of a landscape by the ejecta from the impacts of larger projectiles or by the fragments pro-

duced by collisions between large bodies. With Dione and Rhea it seems likely the small craters were made by the debris from bodies that collided within the Saturnian system. Presumably, therefore, certain regions of Dione and Rhea bear a record of primordial cratering; other regions were resurfaced by material from the interior that buried the oldest craters. The resurfaced regions were then pocked by a second bombardment. We shall refer to the bodies in the second bombardment as Population II.

A problem remains. The "highlands" bearing evidence of cratering by Population I and the plains bearing evidence of cratering by Population II show no correlation with the global pattern of albedo recorded by *Voyager 1* at a distance from Dione and Rhea. On Dione, for example, the leading hemisphere, which was uniformly bright and bland in the distant imagery, turns out at high resolution to have both highlands and plains. What, then, caused the global albedo pattern? Some calculations made by Eugene M. Shoemaker of the U.S. Geological Survey lead to estimates of the contributions that projectiles make today to the cratering of the Saturnian moons. For virtually any source of projectiles outside the Saturnian system the flux of impacts turns out to vary dramatically from the leading hemisphere to the trailing hemisphere of any given moon. For Dione, Shoemaker calculates a variation of 10 to one. For Rhea the variation is six to one. It makes no difference whether the projectiles are comets or asteroids or whether they have periodic orbits around the sun because the gravitational acceleration imparted to an arriving projectile by Saturn overwhelms the projectile's original trajectory.

Shoemaker further calculates that from the apex of Dione to the side of the body (with respect to the orbital motion of Dione around Saturn) the flux of arriving projectiles changes by a factor of

DIONE was photographed by *Voyager 1.* **An image made on November 11, 1980** (*top*), **shows the side of the moon that faces away from Saturn. The orange background is the top of Saturn's clouds, 377,000 kilometers from the moon. Dione is moving toward the right; the trailing hemisphere is at the left. The center of the trailing hemisphere shows bright wisps on a dark field. According to one hypothesis, a pattern of wisps once covered the entire surface of Dione. It was erased by the continual impact of small meteoroids over the history of the solar system, except at the center of the trailing hemisphere, where relatively few such projectiles arrive. An image made on November 12, 1980, at a range of 162,000 kilometers** (*bottom*) **shows what amounts to the prow of Dione. The apex of the leading hemisphere is at about the middle of the terminator along the left of the image. At the right is the side facing Saturn. The beginnings of several bright wisps in the trailing hemisphere are at the right limb of the moon.**

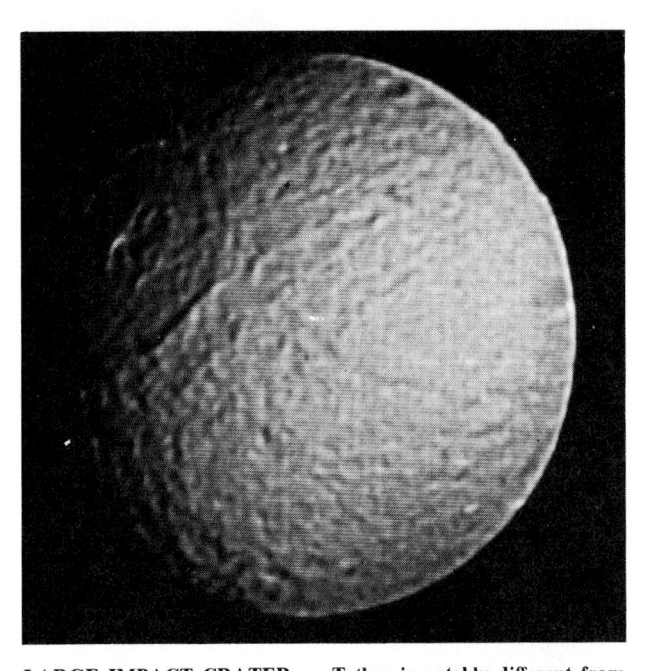

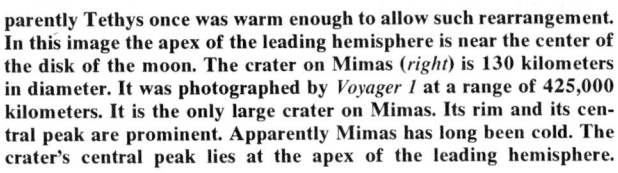

LARGE IMPACT CRATER on Tethys is notably different from one on Mimas. The crater on Tethys (*left*) is more than 400 kilometers in diameter. It was photographed by *Voyager 2* at a range of 826,000 kilometers. When the crater was made, it must have been deep, and its rim and central peak must have been high. Today, however, the floor of the crater has rebounded to match the contour of the moon, and both the rim and the central peak have collapsed. Ap-parently Tethys once was warm enough to allow such rearrangement. In this image the apex of the leading hemisphere is near the center of the disk of the moon. The crater on Mimas (*right*) is 130 kilometers in diameter. It was photographed by *Voyager 1* at a range of 425,000 kilometers. It is the only large crater on Mimas. Its rim and its central peak are prominent. Apparently Mimas has long been cold. The crater's central peak lies at the apex of the leading hemisphere.

only two. From the side to the trailing end it changes by a factor of five. As a result the part of Dione least affected by the continuing impact of projectiles is only a small region at the trailing end of the moon.

With the aid of Shoemaker's calculations a history of Dione can be proposed. In this history the early life of Dione was dominated by the impact of large bodies most likely left over from the accretion of the solar system. Then parts of Dione were resurfaced. Meanwhile collisions near Saturn between the bodies left over from the accretion of the solar system yielded smaller bodies. Some of them took up orbits around Saturn much like Dione's. Their impacts with Dione cratered the newly formed plains moderately. At about this time fractures formed in the surface of Dione. The fractures were filled by bright extrusions from the interior.

It is likely that the entire surface of Dione then had the pattern now seen only at the center of the trailing hemisphere of the body. Over the past four billion years, however, Dione has rotated synchronously and its surface has been "gardened" by the impact of small meteoroids from outside the Saturnian system. The craters made by these bodies may be too small and too scattered to be identified in Voyager images. Still, the craters would rework the surface. In this way they erased the pattern of the surface in a region extending from the apex of the leading hemisphere well into the trailing hemisphere. This history accounts for the global pattern of albedo.

The meteoroids arriving well into the history make up Population III.

Tethys

Tethys is the next moon inward from Dione; it travels in orbit around Saturn at a distance of 5 R_s. Its diameter is almost identical with that of Dione, but its density—1.2 grams per cubic centimeter—is lower. Moreover, its appearance is quite different. *Voyager 1* viewed Tethys only at low resolution. The images showed diffuse patches of small variation in albedo on a heavily cratered surface. The pattern did not resemble the more pronounced global pattern seen on Rhea and Dione. One of the best images showed a branching canyon spanning the distance between the north and south polar regions on the side of the moon facing Saturn. It was estimated the canyon was at least 1,000 kilometers long, 100 kilometers wide and several kilometers deep.

Nine months later *Voyager 2* approached Tethys. The images it transmitted to the earth revealed an enormous impact scar in the leading hemisphere. The rim of the scar has a diameter more than two-fifths the diameter of Tethys itself. As the spacecraft continued its approach the scar was photographed progressively closer to the visible edge of the moon. Soon an image showed it in profile. Here it could be seen that the floor of what must once have been a crater now matches the spherical shape of the body. Only a low rim and a subdued central peak remain.

Evidently the interior of Tethys was sufficiently warm early in the history of the moon to allow the collapse of the raised topography. The same could be said of large craters on the icy Jovian moons Ganymede and Callisto.

Finally the north pole of Tethys and the side facing away from Saturn came into view. From this vantage it was apparent that the canyon found by *Voyager 1* extends over the north pole and down to the equator on the outward-facing hemisphere; thus the canyon traverses three-fourths of a great circle around the body. It could also be seen that parts of Tethys (like parts of Rhea and Dione) had been resurfaced. Smooth plains had developed in a small part of the leading hemisphere, and the resurfacing had buried some of the large craters already there.

A tentative explanation for the current appearance of Tethys begins with the surmise that at the time the large impact scar was made the interior of the body was much warmer and more mobile than it is today. Perhaps it was liquid. If Tethys had been cold and brittle when the scar formed, the impact that formed it might well have fragmented the moon. Moreover, the topography raised by the impact has clearly collapsed. The rim and the central peak of the impact scar persist. Hence it seems likely that much of the crust of Tethys in place at the time of the impact remains at the surface today. One can imagine, then, a simple history in which Tethys freezes from the crust down. If Tethys had first been liquid, the freezing would have increased the volume of the object by about 10 percent and the surface area by about 7 percent. The estimated extent of the canyon on Tethys suggests that it forms 5 to 10 percent of the surface. The canyon may represent the stretching of Tethys' crust over the expanded, frozen interior.

Mimas and Enceladus

The smallest of the nine moons whose presence around Saturn was known before the 20th century are Mimas and Enceladus. Each of them has only a thousandth the mass of a Galilean moon of Jupiter, a thousandth the mass of the earth's moon and a hundred-thousandth the mass of the earth. Their tininess is significant. Before the Voyager spacecraft began to explore the outer solar system the main source of heat in a planet or a moon was taken to be the decay of the radioactive atomic nuclei within it; hence it seemed that the history of a planet or a moon would be determined by the ratio of its volume to its surface area. The volume governs the quantity of radioactive nuclei and thus the generation of heat; the surface area governs the loss of heat and conversely the retention of it. The ratio of volume to surface area is determined by the size of the object: the ratio increases with the radius. One therefore supposes larger bodies are more likely to melt than small ones. Certainly the attribution of internal heat to radioactivity and the resulting correlation of heating with size explains the observations that volcanically the earth's moon is long dead, Mars is moderately active and the earth is quite active. It also suggests, however, that tiny moons such as Mimas and Enceladus should not have evolved substantially since the time they accreted.

The image of Mimas made by *Voyager 1* showed a surface that conforms to this prediction. The surface is uniform in albedo and is saturated with impact craters. It is a surface that has not been reworked by volcanism since the time of the Population I bombardment. In the leading hemisphere of Mimas there is an impact crater 130 kilometers in diameter, about a third the diameter of the little moon. One suspects other craters of that size or even larger should have been made in the course of the bombardment. Why are they not seen? If a cold, brittle Mimas had been blown apart by one such impact, the relative velocities at which the debris would fly apart would probably have been on the order of the velocity required for escape from Mimas' gravity. Such velocities are small compared with the orbital speed of Mimas, which is 30 kilometers per second. Hence the debris would have remained in a narrow band surrounding what had been the orbit of the moon. Gradually the debris would have reaccreted. The result would have been a cold mass of rubble consisting mostly of ice, but with a few percent of rock. (The density of Mimas is 1.2 grams per cubic centimeter.)

The statistics of the craters seen on Mimas suggest this may have happened. On Mimas today the largest crater is the one 130 kilometers in diameter. Then comes a gap in the sizes. The next-largest craters have diameters of only a few tens of kilometers. After that the abundance of craters increases exponentially as the diameter of the craters decreases. It is as if a large-scale impact had destroyed a parent body and the fragments had recombined. The present pattern of craters results, then, from impacts by the last of the fragments. The single largest crater represents a collision not energetic enough to have destroyed the body again; the other large craters made by Population I objects disappeared with the disruption of the parent body.

Voyager 1 provided only distant views of Enceladus, yet if the craters on Enceladus resembled those on Mimas, the im-

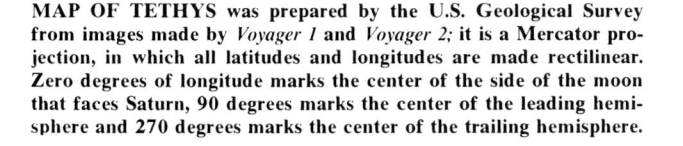

MAP OF TETHYS was prepared by the U.S. Geological Survey from images made by *Voyager 1* and *Voyager 2;* it is a Mercator projection, in which all latitudes and longitudes are made rectilinear. Zero degrees of longitude marks the center of the side of the moon that faces Saturn, 90 degrees marks the center of the leading hemisphere and 270 degrees marks the center of the trailing hemisphere. Two features dominate the topography. One of them is the collapsed crater shown in the photograph at the left on the preceding page. The central peak of the crater lies near 120 degrees longitude and 30 degrees north latitude. The other feature is a great canyon running from south to north between 30 and 330 degrees longitude. The canyon extends over the north pole of Tethys and then back toward the

ages should have shown them. A crater as large as the one in the leading hemisphere of Mimas should certainly have been visible. Instead the images showed a surface that seemed to be smooth. More important, the surface of Enceladus was brighter than that of the neighboring moons Mimas and Tethys. This suggested that much of the surface of Enceladus may have been regenerated and is covered with very fresh ice.

Two chains of reasoning had already implied that Enceladus was unusual. In the first place observations made from the earth had shown that the diffuse ring of Saturn designated the E ring has its maximum brightness along the orbit of Enceladus. This suggested that Enceladus might be the source of the ring. Indeed, it suggested that Enceladus might continually replenish the ring, because otherwise the gravitational field of the moon would tend to clear matter out of its orbit. (It is proposed that such gravitational shepherding is responsible for at least some of the banding discovered by the Voyager spacecraft in the rings of Saturn.)

In the second place, an argument developed by Stanton J. Peale of the University of California at Santa Barbara and his colleagues had led to the conclusion that Io, the innermost major moon

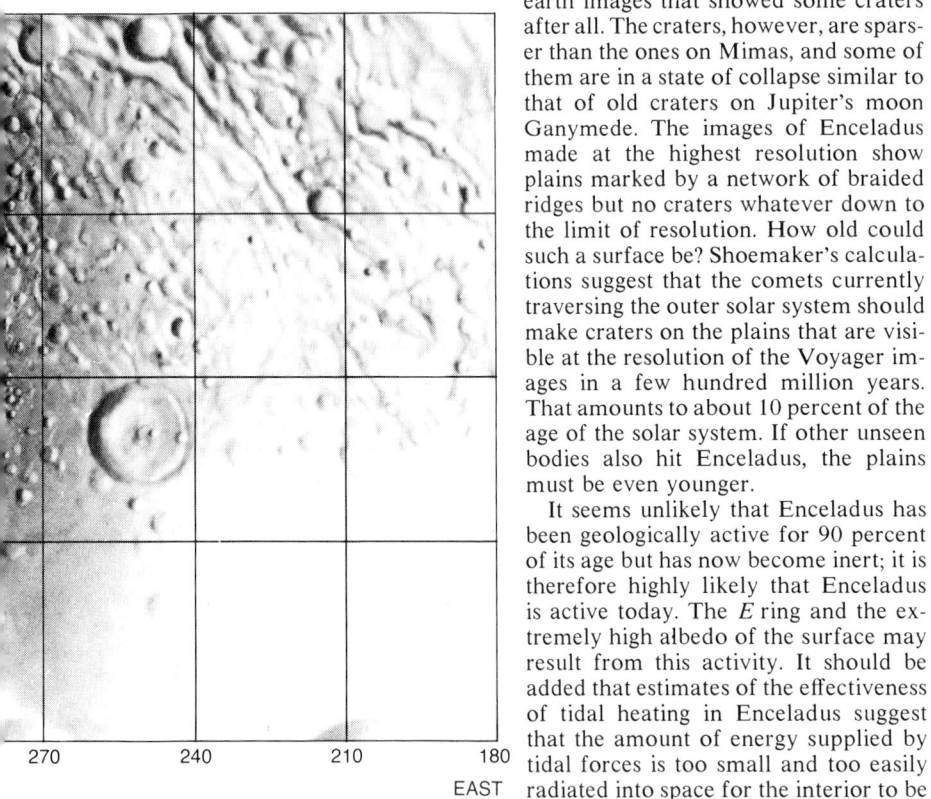

equator; its end is apparent at the upper right of the map. The canyon may represent the stretching of the crust of Tethys over an interior that expanded as it cooled and solidified into ice. The terrain on the map is shown as though it were all illuminated from the west.

270   240   210   180
EAST

of Jupiter, is heated substantially by tidal forces. Then Charles F. Yoder of the Jet Propulsion Laboratory of the California Institute of Technology proposed that a similar mechanism might act on Enceladus. Specifically, the orbital period of Enceladus is half that of Dione. This means that Enceladus and Dione align themselves with Saturn at fixed points in their orbits. The gravitational attraction of Dione for Enceladus acts repeatedly at these points, so that although the attraction is small, it makes the orbit of Enceladus stay slightly elliptical. In such an orbit Enceladus is forced to oscillate radially in Saturn's gravitational field, much as Io oscillates radially in Jupiter's. The internal friction that results from this oscillation might keep the interior of Enceladus warm and mobile. Gases and fluids escaping from the interior might then keep the surface bright and remove the traces of craters. Allan F. Cook II of the Harvard College Observatory and the Smithsonian Astrophysical Observatory and Richard J. Terrile of the Jet Propulsion Laboratory have proposed that if Enceladus is geologically active, gases escaping from the interior of Enceladus and ice particles escaping from the surface might be the source of the E ring's material.

As *Voyager 2* flew close to Enceladus last August it transmitted back to the earth images that showed some craters after all. The craters, however, are sparser than the ones on Mimas, and some of them are in a state of collapse similar to that of old craters on Jupiter's moon Ganymede. The images of Enceladus made at the highest resolution show plains marked by a network of braided ridges but no craters whatever down to the limit of resolution. How old could such a surface be? Shoemaker's calculations suggest that the comets currently traversing the outer solar system should make craters on the plains that are visible at the resolution of the Voyager images in a few hundred million years. That amounts to about 10 percent of the age of the solar system. If other unseen bodies also hit Enceladus, the plains must be even younger.

It seems unlikely that Enceladus has been geologically active for 90 percent of its age but has now become inert; it is therefore highly likely that Enceladus is active today. The E ring and the extremely high albedo of the surface may result from this activity. It should be added that estimates of the effectiveness of tidal heating in Enceladus suggest that the amount of energy supplied by tidal forces is too small and too easily radiated into space for the interior to be a liquid if it consists of water. Perhaps, however, the interior includes methane clathrates or hydrates of ammonia. Such substances have a melting point about 100 degrees Celsius below that of pure water ice.

At least six different types of terrain can be distinguished on the surface of Enceladus. The oldest two are cratered terrains designated $CT1$ and $CT2$. They show a population of craters 10 to 30 kilometers in diameter that almost certainly date back to a period of Population II bombardment in which Enceladus was hit by debris in orbit around Saturn. On the $CT1$ terrains most of the craters have collapsed: their floors have rebounded, their rims have sunk and their central peaks are better described as gentle domes. On the $CT2$ terrains the craters are deep and bowl-shaped. The difference suggests that these different parts of Enceladus have had different thermal histories. The $CP1$ terrains are plains that show a relatively sparse distribution of craters between two and five kilometers in diameter. The craters lie on a rectilinear pattern of faults. Shoemaker's calculations suggest a range of ages. If the flux of recent projectiles has been minimal, the $CP1$ terrains may date back to nearly the time of the Population II bombardment more than four billion years ago. If the flux has been substantially greater, they may be a billion or even two billion years younger. In any event three still more recent terrains ($CP2$, $CP3$ and RP) also formed on Enceladus. They are revealed by the presence of corridors that cut through the older terrains so that parts of craters are obliterated.

The RP terrain includes the plains that show no craters. Its margin is marked by braided ridges. Two tentative hypotheses have been offered for the origin of the ridges. One explanation is that a system of faults was invaded by liquid water that froze. The expansion of the ice is what raised the topography. The other explanation is that a zone of solid-state convection slowly makes matter rise in the center of the plain and sink elsewhere, producing a roughly concentric pattern of wrinkles. Clearly Enceladus joins Io in showing that radioactivity does not always dominate the heating of bodies in the solar system. Enceladus has a hundred-thousandth the mass of the earth. It may nonetheless be just as active geologically.

Different Evolutions

The exploration of the Jovian and the Saturnian systems of moons by the Voyager spacecraft has yielded new insights into the evolution of the small bodies in the solar system. Imagine a three-dimensional chart whose axes are determined by measures of a body's size, composition and thermal evolution. Before the Voyager missions the only bodies that could have appeared in such a chart would have been the planets, moons and asteroids of the inner solar system. All of them fall into essentially the same class of objects of rocky composition, and their degree of thermal ev-

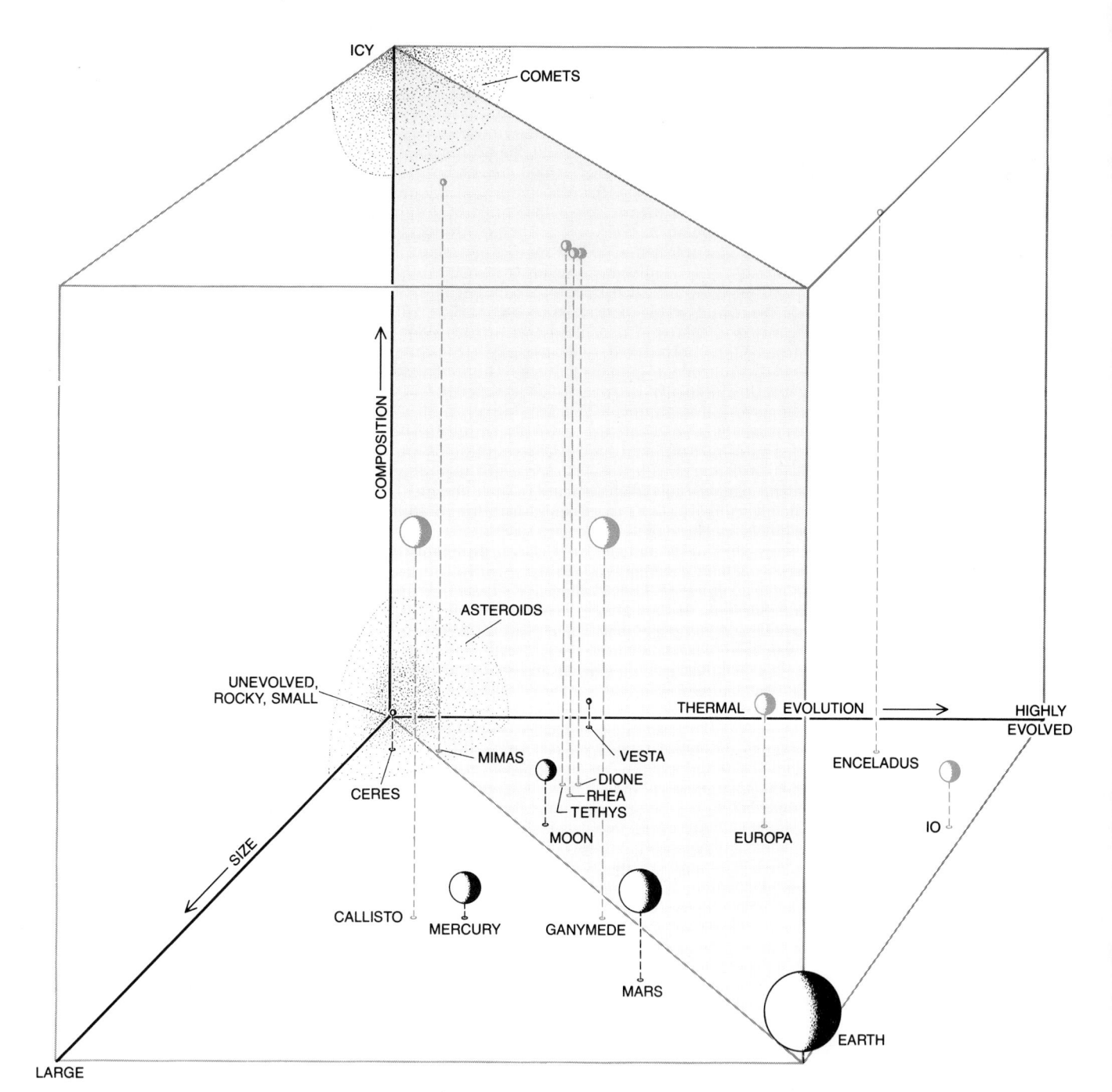

WIDE RANGE OF HISTORIES among planets and moons in the solar system is suggested when planets and moons with a visible solid surface are plotted in a cube whose axes represent their composition (from rocky bodies at the bottom of the cube to icy bodies at the top), their thermal evolution (from cold, unevolved bodies at the left to highly evolved, volcanic bodies at the right) and their size (from small bodies at the back to large bodies at the front). The black spheres are bodies whose position could be plotted before the Voyager spacecraft arrived in the outer solar system. In general these bodies (including Mercury, the earth, the earth's moon and Mars) are rocky. Moreover, the larger they are, the more they show signs of evolution and recent volcanic activity. In the illustration the rocky composition of the bodies means they lie near the bottom of the cube; the correlation between their size and their thermal evolution means they lie near a plane that passes through the cube diagonally from left to right. The correlation between size and evolution implies they derive their internal heat primarily from the decay of radioactive atomic nuclei. The colored spheres represent bodies that can be placed in the cube as a result of the study of their solid surface in Voyager images. Such bodies occupy a large part of the volume of the cube. They include Io and Enceladus, two small but highly evolved bodies that are thought to derive their heat from internal friction as each one moves in a slightly elliptical orbit around its parent planet. Io is a rocky moon, Enceladus is icy. Io, Europa, Ganymede and Callisto are the four major moons of Jupiter discovered by Galileo; Ceres and Vesta are bodies in the asteroid belt between the orbits of Mars and Jupiter.

olution is closely related to their size—a correlation suggesting that radioactive heating has dominated their history. In the wake of the Voyager missions a large part of the chart gains occupants, and the simple trend of evolution with size is destroyed.

Much remains to be discovered about the evolution of the moons in the solar system, but already it is apparent that nonradioactive sources of energy (for example tidal heating) are important for some of them. Moreover, it is apparent that an icy composition can allow vigorous geological activity even in very small moons. Further studies of moons will surely tell us much about both the conditions and processes in the incipient solar system and the ways planets evolve under a wide variety of circumstances.

10

Rings in the Solar System

by James B. Pollack and Jeffrey N. Cuzzi
November, 1981

*Three of the giant planets are now known to have them,
and the rings around Saturn are now known to consist
of myriad ringlets. The form of the rings is maintained
by a complex interplay of sculpturing forces*

The approach of the spacecraft *Voyager 2* to within 100,000 kilometers of Saturn on August 25 marks the climax of a period of planetary exploration in which the rings of Saturn have surprised us perhaps as much as they surprised the first men who saw them almost four centuries ago. Today the rings of Saturn stand revealed in a wealth of detail, including bands, spokes and braids. Some of the detail is unexplained. On the other hand, it emerges that Saturn is not unique in having rings. Jupiter has a ring, and Uranus has at least nine discrete rings. It remains to be determined whether Neptune has rings, and whether rings are therefore ubiquitous among the giant gaseous planets of the outer solar system. Here we shall discuss the structure and composition of the rings of Jupiter, Saturn and Uranus, with special attention to the rings of Saturn. Then we shall take up what is known about the processes that shape the rings. Finally, we shall consider several alternative explanations of how the rings arose.

Strange Appendages

The rings of Saturn were first observed in July, 1610. The observer was Galileo Galilei. Partly because the images produced by his invention, the astronomical telescope, were poor and partly because he had discovered the four largest moons of Jupiter only a few months earlier, he initially thought the blurry, earlike structures he saw were two moons close to Saturn. Soon his opinion changed. For one thing the "strange appendages" did not vary in their position with respect to Saturn from one night to the next. Moreover, in 1612 the appendages disappeared. Today it is plain that the rings had come to lie in an edge-on orientation as viewed from the earth in 1612 and thus had become quite faint.

The geometry of the appendages puzzled astronomers. Indeed, it was proposed that the appendages were handles attached to Saturn or that they consisted of several moons in orbit only around the back of Saturn so that they never cast a shadow on the planet. Finally in 1655 Christiaan Huygens proposed that the appendages were the visible sign of a thin, flat, detached disk of matter arrayed in the equatorial plane of the planet. Depending on the position of Saturn and that of the earth in their orbits about the sun, the disk would vary in its tilt toward the earth; hence it would vary in appearance from a narrow line to a broad ellipse.

For the next two centuries it was assumed that the disk was a continuous sheet of matter. A difficulty with this hypothesis emerged, however, as early as 1675, when Jean Dominique Cassini found that the dark band now known as Cassini's division separates the disk into two concentric rings. Moreover, late in the 18th century Pierre Simon de Laplace showed that the combined forces of the gravity of Saturn and the rotation of the disk would rip apart a single broad sheet of matter. Basically any given parcel of the disk maintains its radial distance from Saturn because two forces are in balance. Gravity pulls the parcel inward; centrifugal force pushes it outward. The centrifugal force arises from rotational velocity; thus the disk is rotating. The problem is that for a rigidly

rotating disk the forces balance at only one radial distance. Laplace therefore proposed that the rings of Saturn consist of many narrow ringlets, each one thin enough to sustain the slight imbalance of forces across its radial width.

The final step toward the modern view of the rings was taken in 1857, when James Clerk Maxwell won the Adams prize of the University of Cambridge for his mathematical demonstration that the ringlets actually consist of numerous tiny masses in independent orbits. Experimental proof for his hypothesis came from several quarters. In 1895, for example, the American astronomers James E. Keeler and William W. Campbell inferred the velocity of the particles in the rings from Doppler shifts: the altered wavelength of spectral lines in the sunlight the particles reflect toward the earth. They found that the rings rotate about Saturn at a rate different from that of the atmosphere of the planet. Moreover, the inner parts of the rings rotate at a greater velocity than the outer parts, just as the laws of physics would lead one to expect for particles in independent orbits.

The rings of Uranus were discovered by accident. A number of groups of astronomers had set out to observe the passage of the star SAO 158687 behind Uranus on March 10, 1977. The intent was to study the structure of the atmosphere of Uranus. Among the more successful observations were those made by James L. Elliot and his associates on the Kuiper Airborne Observatory, an aircraft equipped with a 91-centimeter telescope. His group (and several others) found that the brightness of the star decreased not only when it passed behind Uranus but also at a number of places close to the planet yet far above its atmosphere. The short-duration dimmings defined a series of distances on one side of Uranus that was roughly symmetrical with the series on the other side of the planet. The symmetry was taken to arise from the presence of fairly opaque, quite narrow and nearly circular rings. Later observations have revealed so far the presence of nine rings,

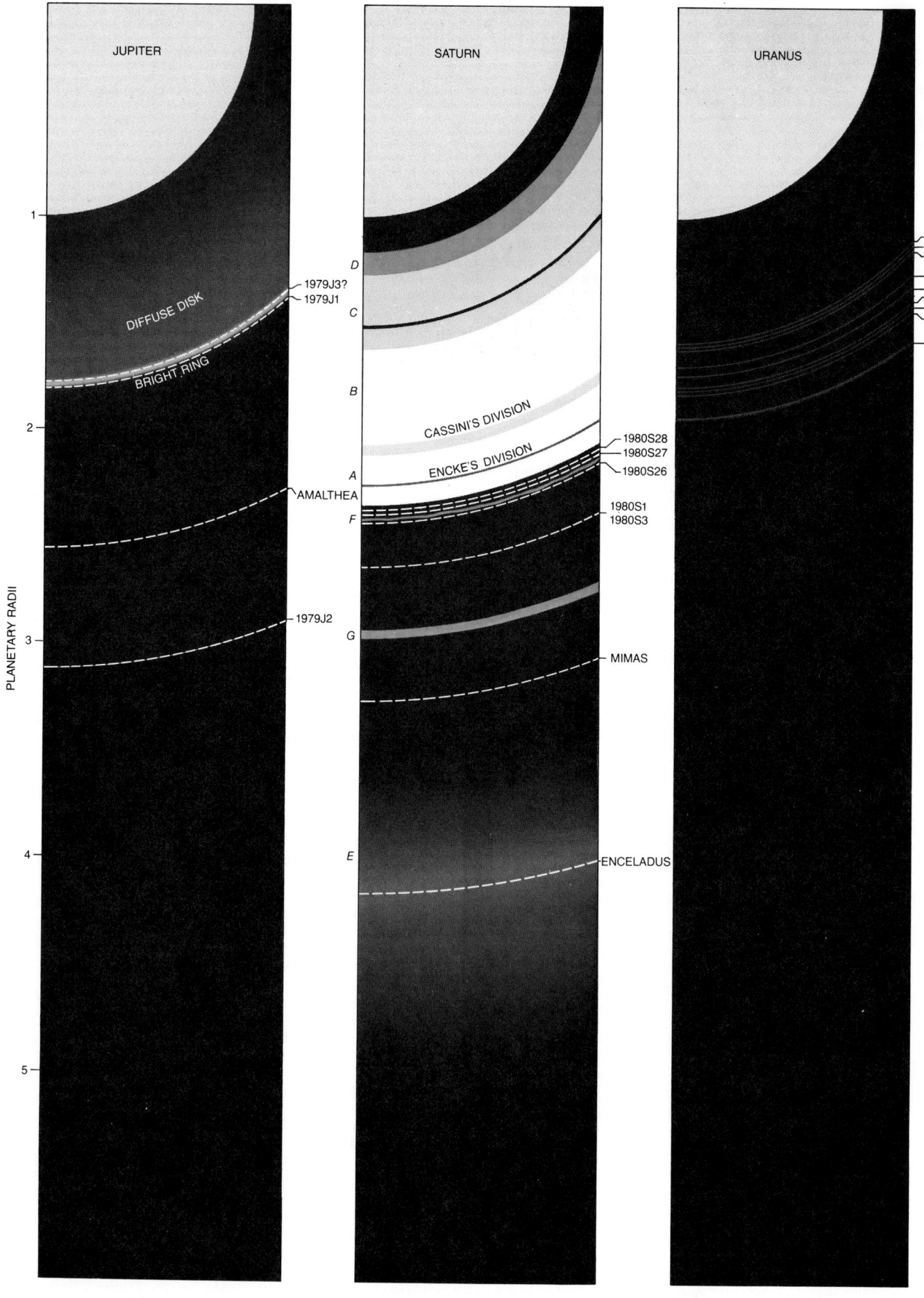

JUPITER

SATURN

URANUS

PLANETARY RADII

1979J3?
1979J1

DIFFUSE DISK

BRIGHT RING

AMALTHEA

1979J2

D

C

B

CASSINI'S DIVISION

ENCKE'S DIVISION

A

F

1980S28
1980S27
1980S26

1980S1
1980S3

G

MIMAS

E

ENCELADUS

6
5
4
α
β
η
γ
δ

ε

all lying within one planetary radius of the top of the Uranian atmosphere.

The rings of Jupiter were discovered by spacecraft. The first hint came when *Pioneer 10* passed near Jupiter in 1974. Jupiter has a magnetic field that traps charged particles in regions of space surrounding the planet. The regions amount to the Jovian equivalent of the earth's Van Allen belt. What *Pioneer 10* detected was a decrease in the count of the high-energy particles in the Jovian belts 50,000 to 55,000 kilometers above the atmosphere of the planet. Mario H. Acuña and Norman F. Ness of the Goddard Space Flight Center of the National Aeronautics and Space Administration suggested that the decrease might be due to the partial absorption of the particles by a close-in satellite or a system of rings. The latter proved to be the case. A faint Jovian ring was detected in 1979 when it was photographed by imaging systems aboard *Voyager 1* and *Voyager 2*. The spacecraft were deflected toward Saturn by the gravitational field of Jupiter. They arrived near Saturn in November, 1980, and August, 1981.

Characteristics of Planetary Rings

The rings of Saturn, Uranus and Jupiter share a number of properties. First, they consist of myriad particles in independent orbits. Second, they lie far closer to their parent planet than the major moons of the planet do; in fact, the bulk of each ring system lies less than one planetary radius away from the surface of the planet. Third, the rings are centered on the equatorial plane of the planet; indeed, almost all the ring material is confined to a thin region in that plane. Fourth, the ring systems of Jupiter and Saturn have a number of tiny

moonlets near or within the rings. It is suspected that similar moonlets are associated with the rings of Uranus.

Still, each ring system has its own peculiarities. The ring system of Saturn has seven major sections. Some of them are separated from neighboring sections by more or less empty annular gaps; the borders of the others are marked by changes in the packing density of ring particles. Each section is designated by a letter that reflects not its distance from Saturn but the order in which the sections were discovered or hypothesized.

The main body of the ring system of Saturn includes, then, the bright and fairly opaque rings *A* and *B*. They are separated from each other by the 5,000-kilometer width of Cassini's division, a rather transparent region but definitely not an empty one. The main body of the Saturnian system also includes the fainter, less opaque *C* ring, which lies inside the inner edge of the *B* ring. It has a degree of opacity comparable to that of Cassini's division. The still fainter *D* ring lies inside the *C* ring. Finally, three very faint rings, *E*, *F* and *G*, lie outside the *A* ring. Taken together, the main rings of Saturn (the *A*, *B* and *C* rings) measure about 275,000 kilometers in annular width, or about three-fourths of the distance from the earth to the moon. In comparison, the thickness of the rings of Saturn is negligible. An upper limit of about a kilometer has been placed on their vertical extent. Compared with their width, the rings are thousands of times thinner than razor-thin.

Information on the composition of the particles in the rings of Saturn can be derived from the rings' ability to reflect or absorb light of different wavelengths. For example, the *A, B* and *C* rings are poor reflectors of sunlight at certain

near-infrared wavelengths. This property is characteristic of water ice, which suggests that water ice is a major constituent of the surface of the particles that make up those rings. Water ice, however, is white in color; that is, water ice is more or less equally reflective at all visible wavelengths. In contrast, the particles of the *A, B* and *C* rings are less reflective in blue light than they are in red light. Perhaps some additional substance is present in small amounts. Dust containing iron oxide has been suggested as the source of the reddish color. It has also been hypothesized that compounds generated by the sun's ultraviolet radiation are responsible for the redness. Certain colorless sulfur-containing compounds become polysulfides under ultraviolet radiation, and polysulfides selectively absorb in the blue. One surprise in the Voyager data is that the particles in the *A* and *B* rings have a similar color but are brighter and more reddish than the particles in the *C* ring and in Cassini's division.

Observations with radar yield further deductions. In 1973 Richard M. Goldstein and Gregory Morris of the Jet Propulsion Laboratory of the California Institute of Technology probed the rings of Saturn with radar waves whose reflection they detected with the 210-foot antenna of the deep-space network at Goldstone, Calif. The high reflectivity of the *A* and *B* rings implied that most of the particles in those two rings are at least comparable in size to the radar wavelengths of several centimeters that the investigators employed. If they had been smaller than the radar wavelengths, they would have been transparent to the radar waves. If they had been much larger than the wavelengths, their emission of thermal radiation at those wavelengths would have been significant. The low level of such radiation limits their size to no more than a few meters.

Data from the Voyager spacecraft have supported and extended the earlier findings. In one type of experiment radio waves from the spacecraft were sent through the rings to the earth and measurements were made of the amount of power scattered by the ring particles at various angles of deflection from the initial path of the waves. As the size of the particles increases with respect to the wavelength, the scattering pattern becomes more narrowly concentrated within small angles of forward deflection. An analysis of the Voyager data by G. Leonard Tyler and Ahmed Essam A. Marouf of Stanford University indicates that the largest abundant particles in the *A, B* and *C* rings are about 10 meters in size. More abundant particles are as small as 10 centimeters, and regional variations in the distribution of sizes are found among the rings.

Just as the scattering of radar waves by the particles in the rings makes it pos-

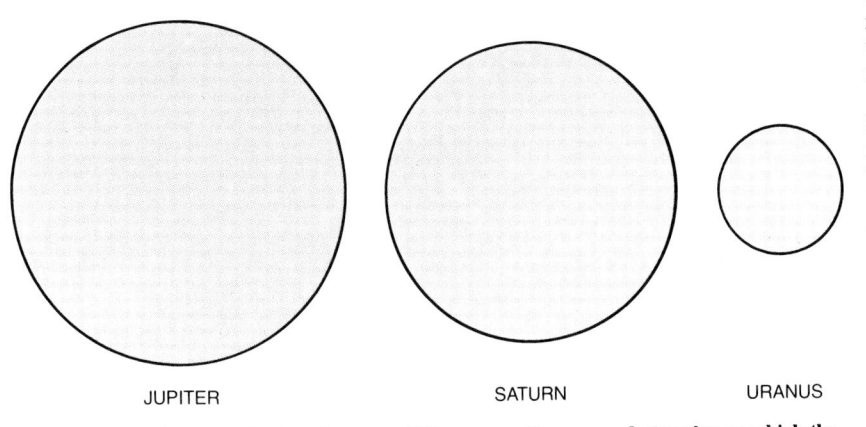

JUPITER   SATURN   URANUS

RING SYSTEMS around Jupiter, Saturn and Uranus are diagrammed at scales on which the radius of each planet is set equal to 1. Jupiter (*left*) has a "bright" ring that is actually quite faint and almost transparent. An even fainter disk of particles extends inward from the ring, perhaps into the atmosphere of the planet. A halo of particles (not shown) gives the system a vertical thickness of some 20,000 kilometers. The system was discovered by spacecraft. Saturn (*center*) has seven rings. The *A* and *B* rings are separated by Cassini's division; the *A* ring includes Encke's division. Letters were assigned to the rings in the order of their discovery. Only the main rings (the *A*, *B* and *C* rings) are readily visible in earth-based telescopes; the rest (except the *E* ring) were discovered by spacecraft. Uranus (*right*) has no fewer than nine rings. Here their width is exaggerated. They were detected from the earth, and they are designated by numbers or Greek letters. The moons in the ring systems are labeled at the right of each panel; their orbits are in broken lines. The relative sizes of Jupiter, Saturn and Uranus are shown above.

SPOKES AND BANDS in the rings of Saturn were first photographed by *Voyager 1*. The spokes arise sporadically in the *B* ring; they are more or less radial wedges. Each spoke is bright in images of the sunlit face of the rings made when the spacecraft and the sun are on opposite sides of Saturn, so that sunlight reflects forward from the ring particles to the spacecraft (*top photograph*), and each spoke is dark in images made when the spacecraft and the sun are on the same side of Saturn, so that sunlight reflects backward (*bottom photograph*). In both images the banding of the rings into ringlets is evident. The wide dark band is Cassini's division.

sible to detect particles on the order of the size of a radar wavelength, so the scattering of sunlight makes it possible to detect particles the size of a wavelength of visible light. In particular, the strong brightening of a segment of ring when it is viewed at small forward scattering angles implies that particles on the order of a micrometer in size are abundant in the segment. Such an observation can be made only when Saturn comes between the sun and the observer. For observations from the earth this condition cannot be met, but for observations from spacecraft it can be. Thus studies of the Voyager data indicate that particles on the order of a micrometer in size constitute a large fraction of the particles in the *F* ring, a significant fraction in many parts of the *B* ring and a less significant fraction in the outer part of the *A* ring. On the other hand, the *C* ring and Cassini's division show no sign of such small particles.

Structure in Saturn's Rings

Before the passage of the Voyager spacecraft near Saturn a limited amount of structure was recognized in the planet's rings. Cassini's division was of course known, and so was Encke's division, a narrower band in the outer part of the *A* ring. The high-resolution photographs of the rings made by the Voyager spacecraft revealed a number of surprises. Narrow annular regions of differing brightness and opacity appeared, seemingly as numerous as the grooves on a phonograph record. In addition deviations from circularity were found. These include radially oriented, wedge-shaped spokes in the *B* ring and knots, braids and twists in the *F* ring.

The greatest amount of detailed structure is exhibited by the *B* ring, which also has the greatest density of particles found anywhere in the rings. Variations in the opacity of the *B* ring occur over radial distances as small as 10 to 50 kilometers. In contrast, the *B* ring has an overall width of 25,000 kilometers. The central, most opaque part of the *B* ring is where the spokes appear. Typically each spoke can be seen for a significant part of the 10 hours it takes a parcel of the *B* ring to complete one orbital revolution. Meanwhile new spokes are arising sporadically in new locations in the ring. Compared with their surroundings the spokes appear bright in forward-scattered light and dark in backscattered light. Hence particles on the order of a micrometer in size are particularly abundant in the spokes.

Each part of a spoke orbits Saturn at the same velocity as that of the ring particles at its radial distance. The inner sections move faster; thus a spoke becomes more tilted with time. Eventually it disappears. The narrow end of each wedge-shaped spoke seems to coincide approximately with the distance from

Saturn at which the period of an orbiting particle matches the period of Saturn's rotation. The magnetic field of Saturn is locked into the planet and therefore rotates with it. Hence electromagnetic forces may be partly responsible for the spokes. In this regard it may be notable that bursts of broad-band static were detected by the Planetary Radio Astronomy experiment aboard *Voyager 1*. The bursts appear to have originated from sources in the *B* ring near the regions of intense spoke activity.

In comparison with the *B* ring the *A* ring is featureless and uniformly bright. It does, however, exhibit a variety of narrow, discrete features in its outermost regions. For one thing the dark and relatively empty band of Encke's division, which has a width of about 350 kilometers, contains two or three narrow, irregular ringlets composed of tiny particles. Images made by *Voyager 2* show that the ringlets have kinks reminiscent of those detected in the *F* ring by *Voyager 1*. Four unusual bands with irregular, wavelike structure bracket Encke's division. At a greater distance from Saturn the outer edge of the *A* ring shows many narrow bands about 20 kilometers wide in which small particles are plentiful. The spacing between the bands decreases outward. The outer edge of the *A* ring is quite abrupt. Here too small particles are plentiful.

The *C* ring and Cassini's division have many structural similarities in addition to their relative lack of redness, their similar degree of transparency and their dearth of small particles. Both of them exhibit discrete, regularly spaced bands of uniform brightness. Both of them also contain a variety of narrow, sharp-edged, completely empty gaps with a radial width of from 50 to 350 kilometers. Furthermore, some of the gaps contain even narrower, equally sharp-edged ringlets that are quite opaque. Several such ringlets are eccentric (that is, noncircular) and nonuniform in width. Images from *Voyager 2* show that the opaque ringlets often differ in color from the surrounding ring material. They more closely resemble the stuff of the *A* and *B* rings.

Rings of Uranus and Jupiter

The rings of Uranus are in many ways the complement of the rings of Saturn. The main rings of Saturn (the *A, B* and *C* rings) are broad, and punctuated by narrow gaps. The rings of Uranus are narrow (they each measure from several kilometers to 100 kilometers in radial extent) and are separated by broad, empty regions. Nine rings around Uranus have been confirmed. Each has been assigned a number or a Greek letter. The technique of stellar occultation exploited to study the rings capitalizes on the minute apparent size of the occulted star to yield a high spatial resolution. Never-

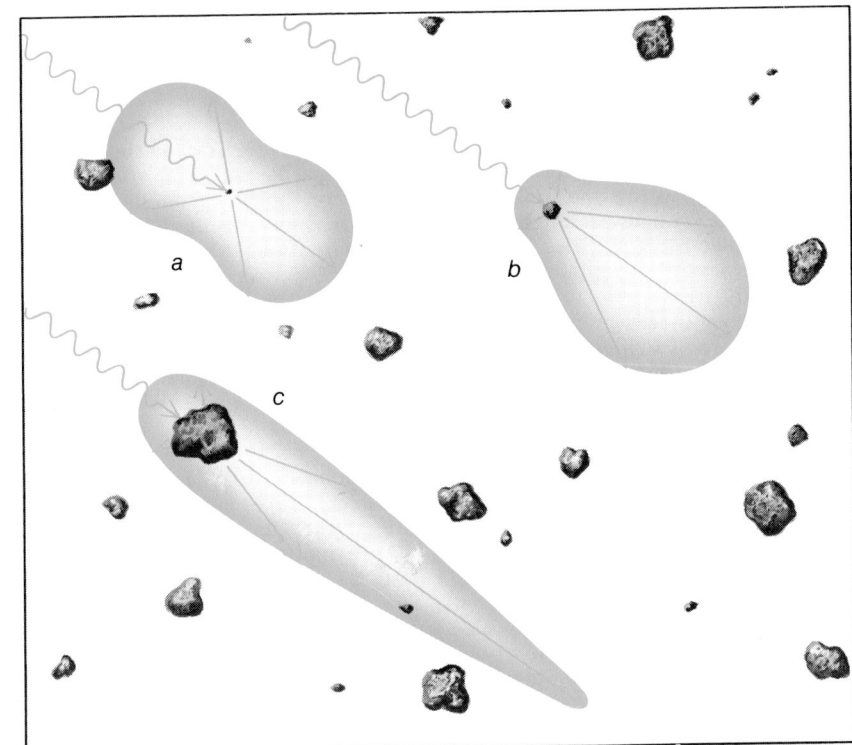

SCATTERING OF SUNLIGHT or some other form of electromagnetic radiation by the particles in a ring allows the deduction of the size of the particles abundant in the ring. In particular, if a particle is less than about a tenth the size of a wavelength of the incoming radiation, it scatters the radiation almost equally in all directions (*a*). If a particle is somewhat smaller than the wavelength, it tends to scatter the radiation forward (*b*). If a particle is larger than the wavelength, it scatters nearly all the radiation forward (*c*). The lengths of the arrows represent the relative amounts of energy scattered at various angles. The observation that particles in the spokes in the *B* ring of Saturn scatter sunlight mostly forward allows the inference that the spokes are local, transient concentrations of ring particles approximately a micrometer in size.

theless, several of the rings of Uranus remain unresolved. Such rings are less than five kilometers in radial extent. (The technique of stellar occultation, as applied to the rings of Saturn by *Voyager 2*, reveals structure on scales as fine as hundreds of meters.) The narrowest rings of Uranus (the ones designated γ, δ and η) are nearly circular and lie in a common plane. In contrast, the somewhat broader rings α, β and 4 tend to be slightly elliptical and inclined to the common plane. The ϵ ring is the broadest and most elliptical of the nine. Its radial width varies linearly with distance from Uranus, beginning at a width of 20 kilometers where the ring is closest to the planet and ending at a width of 100 kilometers where the ring is farthest away. A few of the narrow eccentric ringlets in Saturn's rings are now known to exhibit a similar variation.

The dimming of the brightness of stars as they pass behind the positions of the rings of Uranus indicates that the rings have an opacity comparable to that of the more opaque (and therefore brighter) parts of Saturn's rings. The ϵ ring is particularly opaque. Curiously, it is most opaque at its edges. The gravitational forces arising from the oblate shape of Uranus cause the orientation

of the elliptical rings, including the ϵ ring, to change continuously. A cycle in which the elliptical shape of the ring rotates once about the planet takes some 6,300 hours. In that time a particle in the ring orbits Uranus some 750 times. Hence two things affect the ϵ ring. First, its orientation changes. Second, the particles in the ring at different radial distances from Uranus orbit the planet at somewhat different velocities, in accord with the physics of orbital motion. Still, the ϵ ring maintains its integrity. That is, its eccentricity, its variable width and the concentration of particles toward its edges all persist.

What composes the rings of Uranus? Because of their narrow width the rings are hard to observe from the earth. Nevertheless, recent observations imply that the ring particles are quite dark. Thus water ice is not their dominant component. Quite likely the ring particles orbiting Uranus are made of silicates rich in compounds that absorb sunlight. Certain iron oxides and complex carbon compounds have the required characteristics. Unfortunately nothing is known about the size of the particles.

The rings of Jupiter consist of three main parts: a bright ring, a diffuse disk and a halo. The bright ring has a width

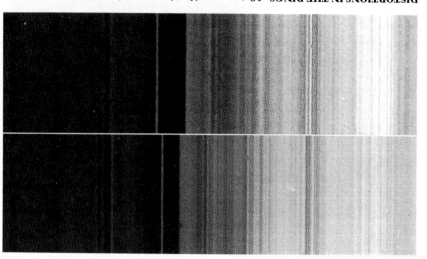

DISTORTIONS IN THE RINGS of Saturn are evident in a comparison of images made on opposite sides of the rings by *Voyager 2*. In each image the *B* ring is the bright region at the left; Cassini's division is the dark region at the right. The outer edge of the *B* ring is shown to be noncircular, and the presence of a bright eccentric ringlet outside the edge of the *B* ring is confirmed. (The ringlet was discovered by *Voyager 1*.) Particles near the outer edge of the *B* ring orbit Saturn twice for each single orbit of the moon Mimas. This resonance greatly amplifies the gravitational effect of Mimas and thereby distorts the *B* ring; overall the resonance raises bulges on opposite sides of the ring. In contrast, the eccentric ringlet has a single maximum that may result from the gravitational field of a local moonlet that has not yet been detected.

of about 6,000 kilometers. Its fairly sharp outer boundary lies about 58,000 kilometers, or .8 Jovian radius, above the surface of Jupiter. In the outer part of the ring a narrow band some 600 kilometers wide is about 10 percent brighter than the rest. Still, the opacity of the "bright" ring is so low that only .001 percent of the sunlight passing through it is intercepted by its particles. The diffuse disk is several times fainter. It extends inward from the inner margin of the bright ring. Indeed, it may intersect the atmosphere of Jupiter. Viewed edge on, the bright ring and the disk appear to be confined primarily to a thickness of no more than 30 kilometers. Remarkably, however, the halo has a vertical extent of some 20,000 kilometers. The halo is highest over the diffuse disk. Its outer boundary extends a little beyond the outer edge of the bright ring.

On the basis of the way the bright ring brightens at small scattering angles it has been found to contain particles whose characteristic size is several micrometers. Such particles are quite ineffective at absorbing high-energy protons and electrons. Hence these particles cannot be responsible for the decrease in the flux of high-energy particles detected in the bright ring by the *Pioneer 10* spacecraft. There must also be particles at least a centimeter in diameter. The particles in the ring are reddish, which makes them the color of many asteroids and moons of the outer solar system.

Collisions in the Rings

The architecture of a ring system results from the interplay of a number of forces. These include gravitational forces due to moons outside the rings and the moonlets embedded in them, and electromagnetic forces due to the planet's rotating magnetic field and even the gentle forces exerted by the dilute gaseous medium in which the rings rotate. All the particles in a ring system share a common orbital motion around a planet: they travel in the direction of the planet's rotation. The vertical and radial motions superimposed on the orbital motion of each such particle have no similar constraint. Hence neighboring particles move randomly in these directions with respect to one another, and collisions are inevitable. When the random relative velocities are large, as they might have been if the rings were ever a thick cloud of particles, the collisions are violent, and even if the collisions are rare, a great amount of the energy of relative motion goes into heating the particles and deforming their structure. The consequent loss of energy means that the random velocities rapidly decrease. The decrease in the vertical component of the velocities leads to a flattening of the ring system. Meanwhile the decrease in the radial component leads to more nearly circular orbits. In brief, a fat ring becomes a thin and roughly circular disk quite early in its history.

Even when the ring particles have lost almost all their random motion, the collisions continue. The reason is as follows. The force of gravity exerted by the planet on the particles in a ring weakens with increasing distance from the planet, so that ring particles at greater distances take longer to circle the planet. Thus a ring particle whose orbit is slightly inside that of a second ring particle eventually catches up with it, and the two of them collide if their radial separation is less than the diameter of a particle.

The collision is likely to happen at a relative velocity of less than a centimeter per second. Nevertheless, it can convert a little of the particles' circular orbital motion into random vertical motion. Subsequent collisions will prevent the particles' vertical velocities from becoming very great. A steady state will therefore be reached that determines the thickness of the ring. If the particles have a wide range of sizes, the smaller particles will gain vertical velocity mostly by being gravitationally deflected in near-miss encounters with the bigger particles. They will lose their vertical velocity mostly by colliding with other small particles. Under these circumstances the small particles will attain a vertical extent several times the size of the biggest abundant particles. In the case of the A and B rings of Saturn the small particles would be expected to occupy a vertical thickness of 10 to 100 meters. Measurements made by experiments on the *Voyager* spacecraft indicate that the main rings of Saturn are certainly no more than a few hundred meters thick. The range of the measurements makes the observational thickness compatible with the predicted thickness.

The collisions of neighboring particles will also convert some of their circular orbital motion into radial motion. Hence the rings will spread radially. An isolated, unconstrained ring will spread until its particles are far enough apart for their collisions to virtually cease. The bright ring of Jupiter may have attained this end state. The width and the very low opacity of the bright ring may reflect the collisionally induced spreading of its larger particles over the age of the solar system.

The rings of Saturn and Uranus, however, display abrupt, well-defined boundaries that limit regions densely filled with particles. Other processes must therefore be counteracting the rapid spreading induced by frequent collisions. An important role in these processes may be played by moonlets embedded in the rings or adjacent to them. The gravitational fields of larger, more distant moons may serve to lock some of the local moonlets into fixed orbits and in that way prevent the moonlets from being dislocated by their gravitational interactions with the ring particles around them.

Fundamentally the motion of bodies in orbit around a much more massive planet is dominated by the planet's gravitational field. In certain instances, however, the gravitational attraction between two relatively tiny orbiting bodies may be amplified and thus come to affect their motion significantly. Such amplification is known as resonant forcing or simply as a resonance. Consider that if the orbital period of a moon is an exact multiple or fraction of that of another moon, the net gravitational effect of each moon on the other is in essence a push or a pull applied repeatedly at the same point in the cyclical motion. Thus the effect is amplified. In some instances resonant forcing locks pairs of moons into orbits whose periods maintain a fixed ratio of two small integers. Such commensurabilities are well known in the Jovian and Saturnian satellite systems, where they involve at least four pairs of the major moons and probably several of the newly discovered moons as well.

Resonance and the Rings

A rather different situation applies to a resonance in a disk of particles. Close to the radial distance from the planet at which the particles in the disk would have an orbital period commensurate with that of one of the planet's moons, the amplification of the gravitational effect of the moon over long periods of time causes the orbits of the particles to become noncircular. Thus the particles are made likely to collide with their less perturbed neighbors. As a result the particles are lost from a band at the radial distance of a resonance. Typically the band has a natural width of some tens of kilometers. Examples of such bands in the rings of Saturn probably include the dozens of narrow gaps in the outer part of the *A* ring, which seem to have resulted from resonances caused by the newly discovered moons designated 1980S1, 1980S3, 1980S26 and 1980S27. The predominance of small particles near the gaps probably attests to the violent collisions induced locally by each resonance.

One additional effect of the resonances that are distant from a moonlet should also be mentioned here. In 1978 Peter M. Goldreich of Cal Tech and Scott D. Tremaine, now at the Institute for Advanced Study in Princeton, proposed that spiral waves of fluctuation in the density of the ring material are present in the rings of Saturn. It had been suggested earlier that similar waves are responsible for the spiral-arm pattern of galaxies such as the Milky Way. In images made by *Voyager 1* spiral density waves seem to be faintly visible in Cassini's division; their tightly wound pattern is reminiscent of a watch spring. It is thought the waves are raised there by resonances. The waves may convey the effects of the resonances over great distances; hence the waves may turn out to play a central role in transporting material within the disk formed by Saturn's rings.

The resonances created in a disk by a moon are spaced closer together as the orbit of the moon is approached. At some critical distance the radial spacing between successive resonances becomes equal to the natural width of each resonance. Within this critical distance the resonances overlap. The result is a continuous zone of transfer of ring material that clears the ring material away from the orbit of the moon. The width of the zone and the degree of clearing in it will depend on the mass of the moon and the density of the surrounding ring material. The more massive the moon, the wider the zone, but the denser the packing of ring particles, the more often it will happen that collisions between particles will propel some of the particles back into the zone. Jack Lissauer and Frank H. Shu of the University of California at Berkeley and M. Hénon of the Nice Observatory have shown that if moonlets from several kilometers to tens of kilometers in size were embedded in the rings of Saturn, the moonlets could cause much of the fine structure in the rings. As of this writing, however, no moonlets have been detected in even the most likely locations in the images made by *Voyager 1* and *Voyager 2*.

If the moon is adjacent to a ring, still another effect is possible: the overlapping resonances that surround the orbit of the moon can prevent the ring from spreading and give the ring a sharp boundary. The outer edge of Saturn's *A* ring is probably maintained in this way by the moons 1980S26 and 1980S27. The ring in turn repels the moons. As it happens 1980S26 is quite close to a radial distance from Saturn that would trap it in a resonance with the much more massive moon Mimas or Tethys. Such a resonance would "anchor" its radial distance, so that it could continue to sculpture the outer edge of the *A* ring. The only trouble with this hypothesis is that precise measurements seem to place

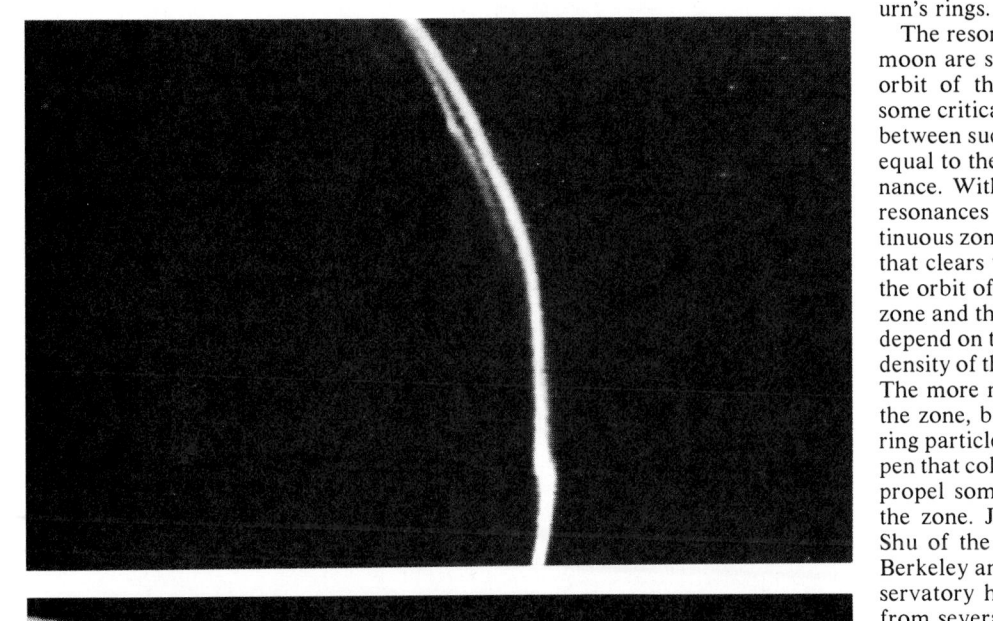

BRAIDS IN THE "F" RING of Saturn were revealed in photographs made by the *Voyager 1* spacecraft when it passed near the planet in November, 1980, but when *Voyager 2* arrived nine months later, they were gone. The ring itself is some 80,000 kilometers (1.3 Saturnian radii) from the surface of Saturn and 4,000 kilometers from the outer edge of the *A* ring. In November, 1980 (*top*), it consisted of three strands, each some 30 kilometers wide. The outer two strands exhibited kinks, warps and knots, and they seemed to be braided, or at least to intersect. The gravitational fields of the moons designated 1980S26 and 1980S27 may account for such structural distortion. By August, 1981 (*bottom*), the *F* ring had changed its structure substantially. An unbraided strand was dominant; three fainter strands appeared as companions.

1980S26 slightly away from the anchoring distance.

Finally, two moons whose orbits are close together can prevent a narrow ring between their orbits from spreading radially. For example, 1980S26 and 1980S27, whose orbits are separated radially by only some 2,000 kilometers, may act as shepherds to confine the narrow, multistrand F ring between them. When one or both of the moons pass close to a given strand of the ring, their gravitational tug may generate distortions such as the ones seen in the images of the ring made by *Voyager 1*. Still, it is not clear why the F ring should consist of multiple narrow strands rather than a single somewhat broader strand, and why one of the strands recorded by *Voyager 1* was untwisted. In images made by *Voyager 2* the F ring exhibits no braids or other distortions, and the number of strands seems to have changed.

Whether shepherding moonlets produce smooth ringlets (such as the narrow, eccentric ringlets in Saturn's C ring and Cassini's division and probably the ε ring of Uranus) or kinky ringlets (such as those of Saturn's F ring and the ones in Encke's division that were found to be kinky in images made by *Voyager 2*) may depend on the ringlets' degree of opacity, or equivalently their particle density. In a ringlet that is rather transparent collisions between particles are relatively infrequent; hence the "echoes" of past

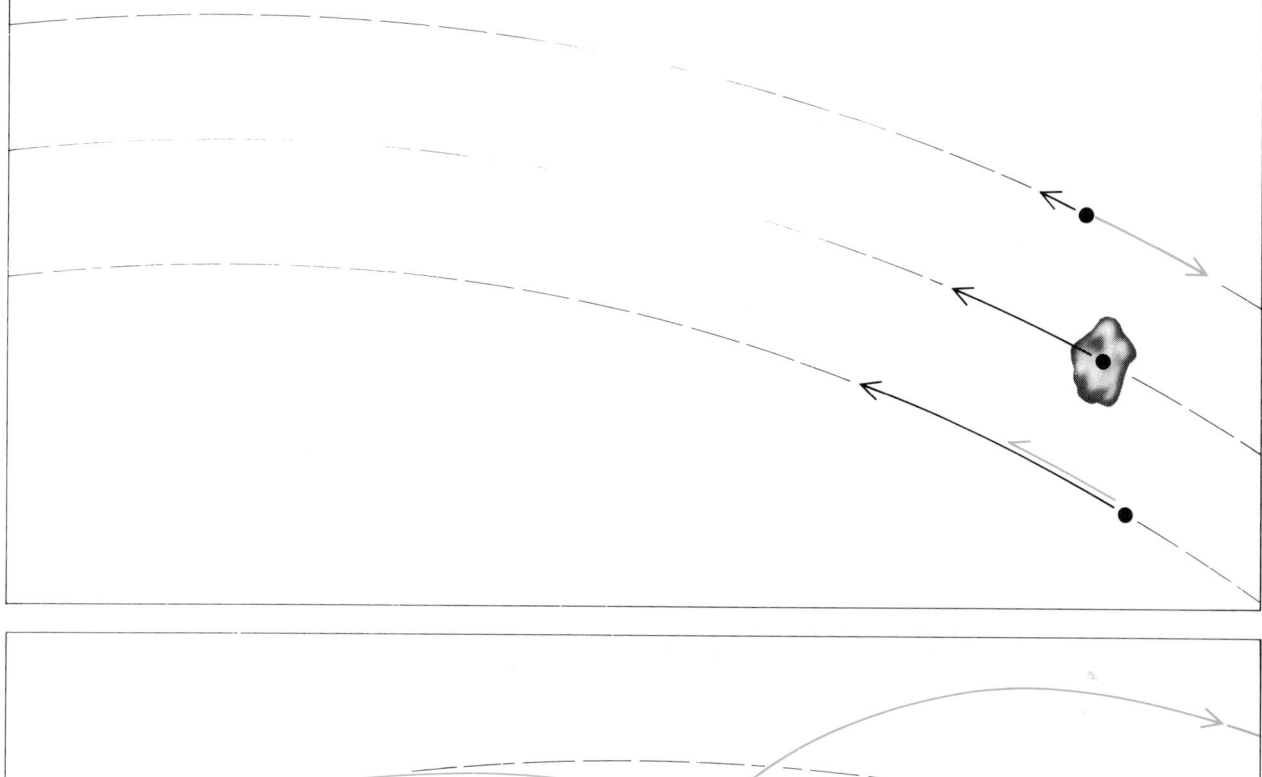

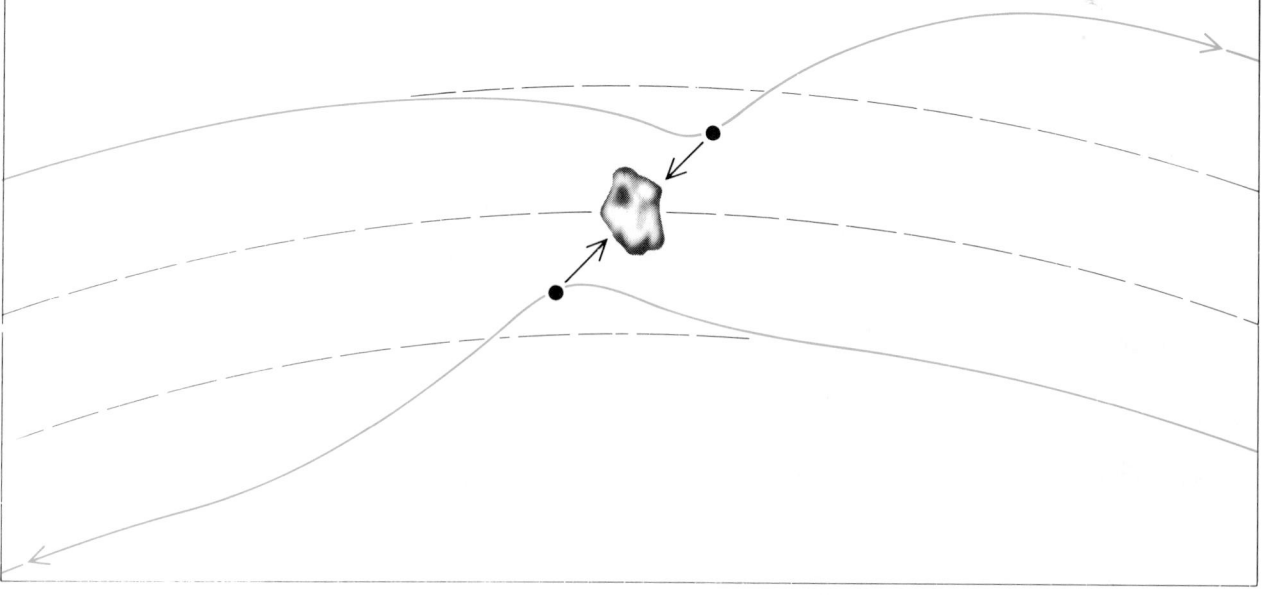

GRAVITATIONAL SHEPHERDING of particles by moonlets in the midst of a ring system may account for some of the banding in the rings of Saturn. The top drawing shows a moonlet and two representative particles in orbit around a planet. In accord with physical law the inner particle moves faster than the moonlet, which in turn moves faster than the outer particle (*black arrows*). Thus the moonlet is being overtaken by the inner particle at the same time as it in turn overtakes the outer particle (*colored arrows*). Each particle is drawn toward the moonlet by gravitational attraction; hence each particle is closer to the moonlet just after they are neck and neck in their orbits around the planet than it was just before. The bottom drawing shows the result. The net gravitational tug the moonlet exerts on the outer particle is in the direction of the outer particle's orbital motion, and so the outer particle is raised to a higher orbit. Conversely, the tug the moonlet exerts on the inner particle is opposite to the direction of orbital motion; the inner particle falls into a lower orbit. Ultimately the moonlet clears out a band surrounding its trajectory. The more massive the moonlet is, the wider the band will be.

gravitational perturbations may persist. In a ringlet that is more nearly opaque the greater incidence of collisions may damp the perturbations.

A full understanding of the physics of the interactions of rings with moonlets is not yet at hand. For example, Stanley F. Dermott and Thomas Gold of Cornell University have offered a hypothesis in which a narrow, eccentric ringlet with sharp edges is maintained by a single small moonlet hidden inside it rather than by a pair of surrounding moons or moonlets. Moreover, the lack of success (so far) at finding moonlets embedded in the main rings of Saturn means that other explanations of the fine radial structure in places such as the *B* ring must be explored. One hypothesis proposes that ring particles may respond to transient increases or decreases in their packing density by a further increase or decrease. Suppose the density of particles increases locally. The collisions between the particles become more numerous, and since some of the energy of the collisions is lost in the heating and deforming of the particles, their energy of random motion decreases. Unless the further collisions at lower velocities lose a smaller proportion of their energy the process of collision and coalescence continues. It is proposed that the overall

result might be radial variations in the opacity of a ring.

Other Forces in the Rings

Small particles in the rings can be affected by forces other than gravity. Electromagnetic forces are an important example. The rings of Jupiter, of Saturn and possibly of Uranus lie in the midst of a plasma of low density, that is, they lie in a tenuous gas consisting of negatively charged electrons and positively charged ions. The electrons are less massive than the ions; therefore they move faster, and initially they collide with particles in the ring more often than the ions do. Eventually the particles become negatively charged by their absorption of electrons. At that point their charge repels the arrival of further negatively charged particles. More important, the ring particles themselves are now accelerated by an electromagnetic force as they cross the planet's magnetic field. If the particles are smaller than about .1 micrometer, the electromagnetic force is greater than the planet's gravitational attraction, and so it dominates their motion.

Several aspects of the structure of ring systems might thereby be explained. With Jupiter, for example, the axis of

the magnetic field is tilted by some 10 degrees with respect to the axis of the planet's rotation. In such circumstances electromagnetic forces can give small particles a vertical distribution much greater than that of larger particles. The vertical extent of the halo of Jupiter's ring system is comparable to the extent that would be expected for particles whose size is .1 micrometer or smaller. With Saturn electromagnetic forces may turn out to be important for the peculiar structure of the *F* ring. Moreover, a number of ingenious theories have been advanced to explain how the spokes in Saturn's *B* ring arise without a simultaneous disturbance of the finely banded radial structure of the ring. Some of these theories invoke a rain of charged particles from either the planet or the rings themselves as a way of transferring electric charge to micrometer-size particles and lifting them from the surface of larger particles. The lifted particles are recaptured when they later collide with other large particles.

A second force that may dominate the motion of small particles in the rings is gaseous drag. Here the friction due to the presence of the plasma causes ring particles to spiral toward the planet. The smaller the particle, the faster the decay of its orbit. In only some 20 years, for example, gaseous drag can cause a micrometer-size particle to move from the outer edge of the bright ring of Jupiter to the inner edge. Another 200 years would take it through the diffuse disk and into the planet's atmosphere. The relatively structureless appearance of Jupiter's bright ring may arise in part from this rapid radial motion of the small particles in it. Indeed, the diffuse disk may be populated by particles that move into it from the bright ring.

In addition to gravity, electromagnetic forces and gaseous-drag forces, the small particles in a ring system are subjected to collisions that destroy them. For one thing interplanetary space contains solid bodies whose size ranges from less than a micrometer to more than a kilometer. The smaller bodies are by far the most numerous. The interplanetary bodies are in orbit around the sun, whereas the ring material is in orbit around a planet. Hence the two can collide at velocities of several tens of kilometers per second. If the interplanetary body is larger than about a hundredth the size of the ring particle, the collision destroys the particle. On that basis it can be estimated that a micrometer-size particle in any of the ring systems lasts for only about 10,000 years. Another type of collision may be even more destructive. In Jupiter's ring and in the outer rings of Saturn it is likely that a particle smaller than 10 micrometers is eroded by its collisions with high-energy ions in the planet's Van Allen belt well before a micrometeoroid hits it.

How did the rings arise? The sun, the

VARIOUS DEGREES OF TRANSPARENCY and thus of the packing density of particles in the rings of Saturn are evident in an image made by *Voyager 1* that shows the unlit face of the rings as they cross the bright disk of the planet. The disk shines through most of the *C* ring (*bottom*), and it shines through Cassini's division to a similar extent, although both include ringlets whose opacity is greater. In contrast, much of the *B* ring (*middle*) is almost opaque.

planets, the moons, the asteroids and the comets of the solar system are all thought to have formed about 4.6 billion years ago inside the solar nebula, a diffuse cloud of gas and dust. The sun and the giant planets Jupiter, Saturn, Uranus and Neptune formed in large part (or even entirely) from the gases of the nebula; Mercury, Venus, the earth, Mars, Pluto, the moons, the asteroids and the comets formed from solidified matter. The composition of the solidified matter would have depended on the temperature of the nebula from place to place. In general, however, the solidified matter consisted of varying mixtures of "rock" (which included silicates and iron) and "ices" (which included water).

Circumplanetary Condensation

At first Jupiter and Saturn, and perhaps Uranus and Neptune, were several hundred times larger than their current dimensions. Under the influence of their own gravitation they gradually shrank. As they did so they rotated faster. Eventually the increasing centrifugal force associated with the rotation caused the outermost part of each incipient giant planet—in essence a gaseous envelope—to become distinct from the more nearly spherical concentration of gas inside it, which continued to contract. The envelope served as a source of the solid grains from which the large moons of the planet accreted. Some of the grains would have formed from the condensation of gases such as water vapor as the envelope cooled. Perhaps the same process of accretion of grains into larger objects gave rise to a ring system somewhat closer to the planet.

After an interval on the order of a million years the circumplanetary envelopes disappeared. Perhaps they were blown away by a wind of ionized gas from the youthful sun, or perhaps friction within the envelopes caused them to collapse onto their planets. In any case the formation of moons, moonlets and ring particles ended.

Today Jupiter and Saturn give off about twice as much energy to space in the form of infrared radiation as the amount of energy they absorb in the form of sunlight. The excess represents the conversion of gravitational energy into heat by each planet's past and present contraction. At the time their moons were forming the planets were contracting much more rapidly. Hence Jupiter and Saturn and perhaps the other giant planets may have radiated many thousands of times more heat then than they do now. This heat may well have controlled the temperature and thus the composition of the solid matter in the circumplanetary envelopes. At a given time the temperature would have increased toward the planet and at a given place the envelope

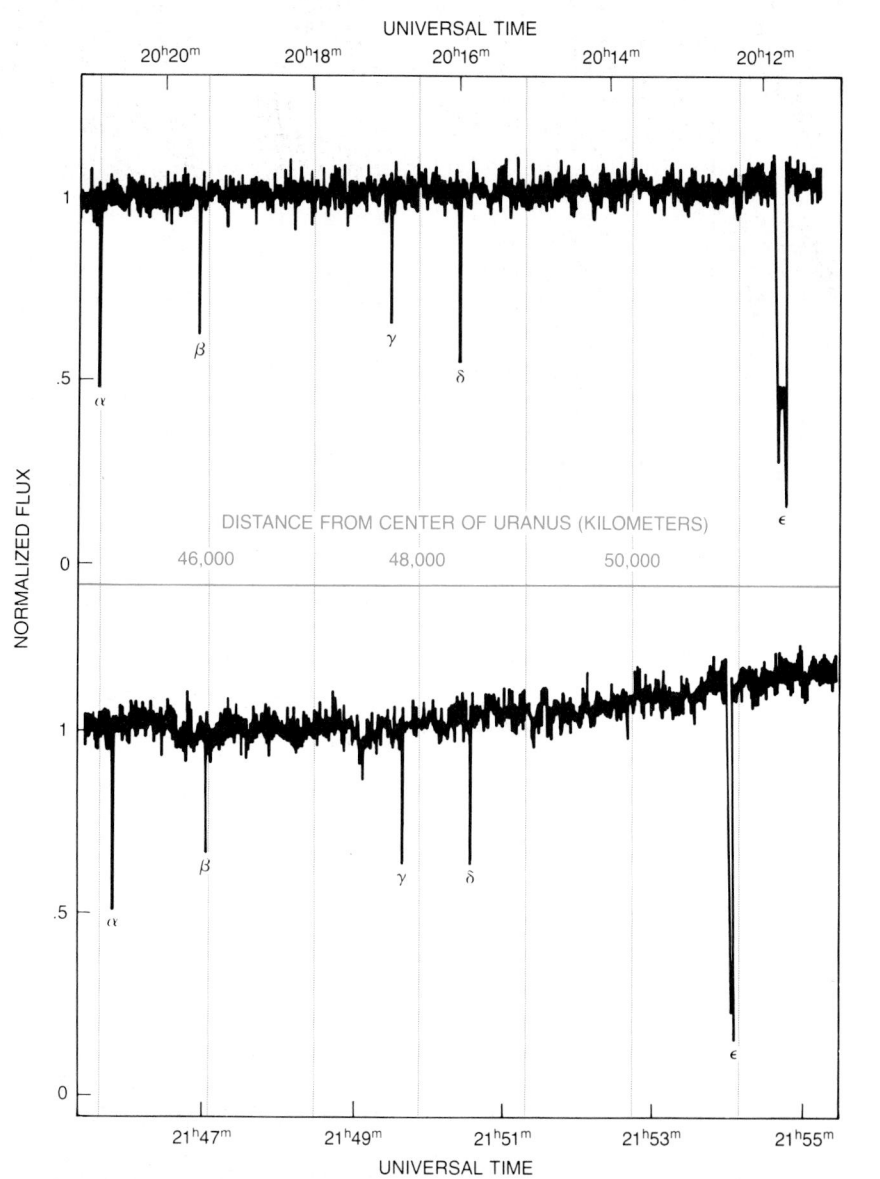

URANUS' RINGS WERE INFERRED from the series of dimmings recorded as a star passed behind Uranus in 1977. The pattern of dimmings minutes before the star disappeared behind the planet (*top*) matched the pattern minutes after it reemerged (*bottom*). For four of the dimmings the match was almost exact; thus four rings are nearly circular. A fifth ring, designated epsilon, is notably eccentric. Further observations have shown four more rings. The data were collected by James L. Elliot and his colleagues on the Kuiper Airborne Observatory.

would have steadily cooled with time.

Several hypotheses about the origin of the ring systems around three of the giant gaseous planets posit that ring material formed in the circumplanetary envelopes at positions close to its current radial distances. Variations in temperature and other conditions near each planet may thus have been the cause of striking differences in the composition of their close-in moons and their ring material. The mean densities of the four largest moons of Jupiter imply that the inner two consist almost exclusively of rock and the outer two of comparable amounts of ice and rock. Since Jupiter's ring is well inside the orbit of the innermost large moon, any particles that formed near the ring could have consisted only of rock. Indeed, they

could have consisted only of the relatively rare minerals that condense at high temperatures.

In contrast, the innermost large moons of Saturn all consist substantially of ice. Thus the temperatures even at the small distance from Saturn where the planet's rings now lie may have been cool enough to allow the formation of particles made up in large measure of water ice. The differences between the conditions around Jupiter and those around Saturn are readily explained. First of all, the amount of heat generated in early times by the contraction of a gaseous sphere depends approximately on the square of its mass. Jupiter, with more than three times the mass of Saturn, emitted 10 times more heat. It may also be that not much rock was available near Saturn for incorporation into particles. When Saturn's circumplanetary envelope formed, the planet's atmosphere probably extended into the region now occupied by the rings and the innermost moons. Hence most of the rocky grains there may have been kept in Saturn's atmosphere. They may also have been claimed by the moons farther out.

The hypotheses that the temperature near Saturn was relatively low and that little rock was available there are consistent with the finding that the inner moons of Saturn have a lower mean density and consequently a larger fractional content of ice than most of the outer moons. The hypotheses may also help to explain why Saturn's rings seem to be almost pure water ice, whereas Jupiter's are rock. Yet it seems the rings of Uranus are rock. One would think they would have been ice.

Rival Hypotheses

The various hypotheses that explain rings around the giant gaseous planets differ primarily in their account of the relation between the ring particles observed today and the primordial ring

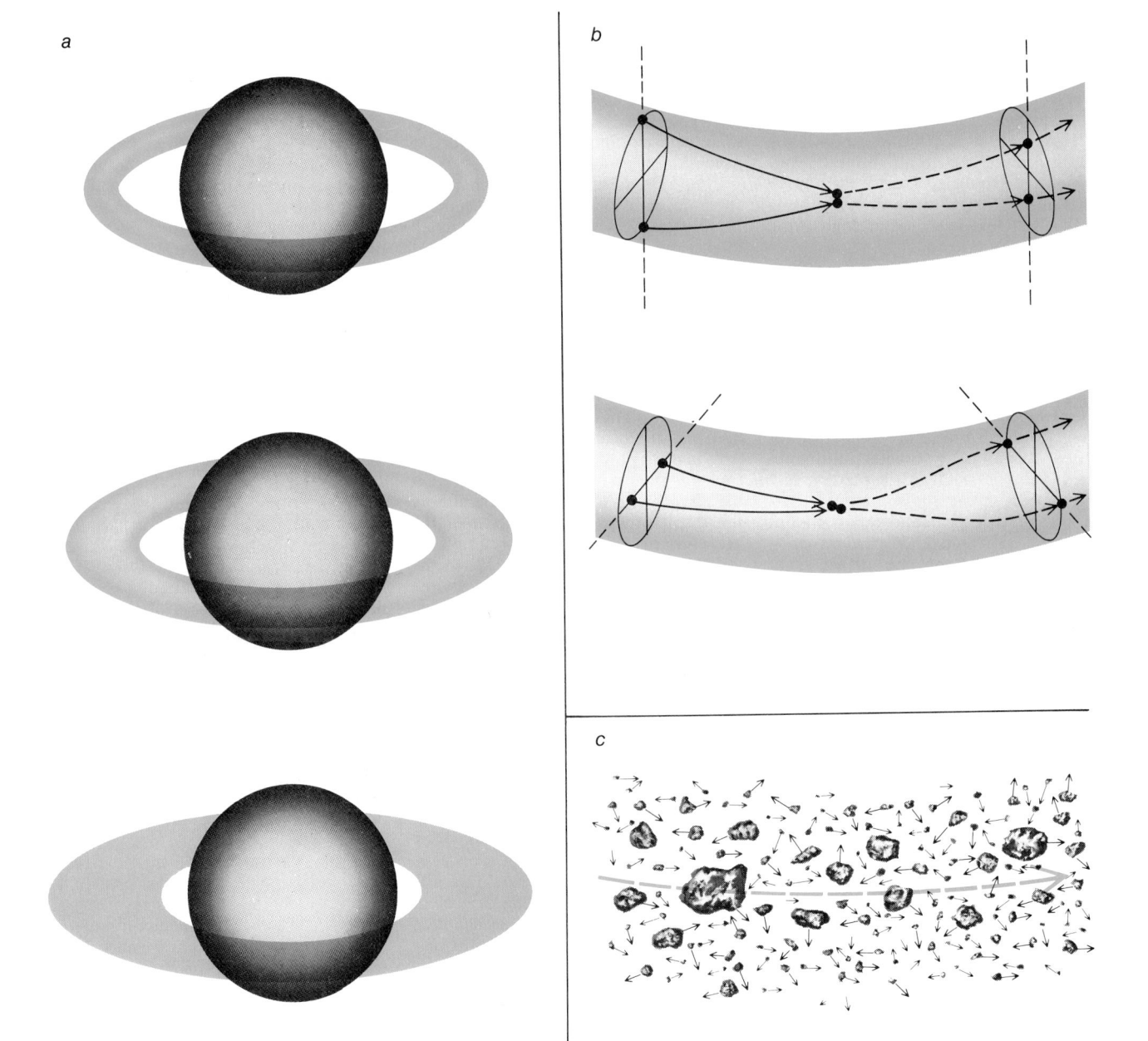

EVOLUTION OF A RING SYSTEM probably began in the early solar system, when the contraction of a rotating gaseous cloud into a giant planet such as Saturn left behind a gaseous envelope strung out by the centrifugal force associated with its rapid rotation (*a*). The particles accreting in the envelope must have orbited the planet, but they also had random motions that caused them to collide (*b*). The collisions were inelastic: they deformed the particles and also heated them. The corresponding loss of kinetic energy decreased their vertical motion. The collisions between two particles at nearby radial distances had an additional effect. In each such collision the inner particle would have been orbiting the planet faster; it would hit the outer particle from behind. The resulting transfer of momentum would raise the orbit of the outer particle and lower the orbit of the inner one. Thus the envelope spread horizontally. Ultimately a flat disk emerged in which the random velocities of the particles with respect to one another had only about a ten-millionth the magnitude of their mean orbital velocity (*c*). It is thought the smallest particles in the rings today result from the collisional erosion of the larger particles.

material. According to one hypothesis, a single large body was fragmented into myriad pieces when it came close to a planet and the fragments then formed rings. The body may have been a large meteoroid that suffered a chance gravitational encounter with the planet, or it may have been a moonlet that formed in the planet's envelope. In either case the agent of fragmentation would have been tidal disruption: the shearing force that arises because the gravitational attraction exerted by the planet on the body is greater for the parts of the body closer to the planet than it is for the parts farther away. The creation of Saturn's rings by tidal disruption was first proposed by the French mathematician Edouard Albert Roche in 1848. Roche had calculated that tidal forces exceed the cohesive self-gravitation of a liquid moon if the moon comes closer than about 1.5 Saturnian radii to the surface of the planet. This disruption threshold—the Roche limit—lies close to the outer edge of the main rings of Saturn.

It is quite unlikely, however, that a moonlet near Saturn would have been liquid. It would have been solid, and a solid moonlet is held together not only by self-gravitation but also by the forces that order the atoms in crystalline matter. According to Hans R. Aggarwal and Verne Oberbeck of the Ames Research Center of NASA, a solid moonlet smaller than about 100 kilometers in diameter cannot be tidally disrupted at any distance from the surface of a planet. Moreover, a larger solid moon cannot be disrupted at a distance greater than .4 planetary radius from the surface. That distance places the disruption threshold inside the inner edge of the main rings of Saturn. It also is unlikely that the tidal disruption of a stray meteoroid could yield ring particles near Saturn. The particles, like their parent body, would have velocities sufficient for them to escape from the planet's vicinity.

There is nonetheless a subtle way in which tidal forces could have been important. As Roman Smoluchowski of the University of Texas at Austin has shown, the gravitational attraction of particles of equal size is insufficient to hold them together against disruption by tidal forces inside the classical Roche limit (the limit calculated by Roche for a liquid moon). In contrast, two particles that differ greatly in size can resist tidal disruption at distances well inside the classical limit. As it happens, the limit set by the tidal disruption of two particles of equal size and the limit set by the tidal disruption of two particles of unequal size lie close to the outer and inner edges of the ring systems around Jupiter, Saturn and Uranus. Within the outer disruption limit particles may have accreted, but so slowly that they were kept from aggregating into large moons. Within the inner disruption limit growth may have been almost impossible.

The second major hypothesis regarding the history of ring particles was suggested by Eugene Shoemaker of the U.S. Geological Survey. It postulates that a single large moon in the ring region (or perhaps a number of moons) collided catastrophically with a stray meteoroid. The photographs of the moons of Jupiter and Saturn made by the Voyager spacecraft do in fact show that the moons are scarred by a large number of craters produced by high-velocity collisions. Mimas, one of the smaller moons of Saturn, has a crater spanning a third of one hemisphere. The collision that made this crater must have come close to destroying Mimas. Nearer Saturn two objects with a diameter of about 100 kilometers occupy almost identical orbits. They may be the largest fragments of a catastrophic collision between a moon and a giant meteoroid.

There are several reasons why catastrophic collisions may have occurred preferentially in a region to be occupied later by rings. First, the major moons of Jupiter and Saturn tend to be smaller at distances closer to their planet. At a given energy of collision small moons are more likely to fragment than large ones. Second, the gravitational field of a planet focuses the trajectories of meteoroids, so that the flux of meteoroids passing close to the planet is significantly greater than the flux at increasing distances.

A final hypothesis regarding the history of ring particles postulates that the larger bodies in the rings are simply the result of the limited extent of the accretion of matter in the circumplanetary envelope at distances close to the planet. The accretion began with the cooling of the envelope and the resulting condensation of gaseous matter into tiny solid grains. Gravitational forces and gaseous drag caused the grains to settle into the equatorial plane of the envelope. There the grains continued to grow by the condensation of vapor onto their surface. Such growth could bring them to sizes as large as a few meters. The particles that make up most of the visible rings of Saturn range in size from a few centimeters to a few meters. They may be the product of such a process. Any moonlets in the rings would then be the local products of a stage of growth in which the meter-size bodies accreted further as the result of gentle collisions.

The ring particles in all three ring systems are small and numerous. According to the accretional hypothesis, there are several reasons why this is the case. First, the undisturbed formation of tiny grains could not begin at a given distance from an incipient giant planet until the planet had shrunk to a size inside that distance and the circumplanetary envelope had cooled sufficiently. Thus less time was available for the formation of grains near the planets than was available at greater distances. Second, the tiny grains near Jupiter could form only from the relatively rare, high-temperature condensates that were available. Finally, the arrival of a moonlet at a certain size would mean that its overlapping resonances spanned a width comparable to its dimension. Fresh material could no longer reach it, and so it would stop growing. In Saturn's rings the calculated limit to growth is a few kilometers to some tens of kilometers. Rather different conditions must have prevailed at greater distances from Saturn and the other giant planets, since bigger moons formed there.

Continuing Erosion

What about the small particles in the ring systems? We have already noted that gaseous drag causes micrometer-size particles to spiral from the outer edge of the bright ring into the atmosphere of Jupiter in a mere few hundred years. Plainly such particles could not have survived from the time the planet had a gaseous envelope. They must be forming today. According to Joseph A. Burns of Cornell, they result from the erosion of larger bodies in or near Jupiter's bright ring.

Ring particles larger than a centimeter or so are not readily destroyed by their collisions with interplanetary micrometeoroids. Instead each collision excavates a tiny crater around the point of impact, and an amount of matter 1,000 to 10,000 times greater than the mass of the impacting body is ejected. Many of the micrometer-size particles in the ring systems may originate, then, as ejecta. It can be estimated that if a moonlet is smaller than about 10 kilometers in diameter, most of the ejecta resulting from a collision with an interplanetary body can escape from the moonlet's gravitational field. The ejecta that escape from the moonlet would lack the energy to escape from the planet around which the moonlet orbits; thus they would take up orbits in the rings. (The continuing erosion of a parent population might also be a source of bodies in Saturn's rings on the order of a centimeter or a meter in size.)

In sum, the moonlets in the rings and also the largest ring particles probably date back to the early history of the solar system: they are contemporaries of the moons of the giant planets. The smallest ring particles are forming even now. It is suspected that the large moons of the outer solar system and also several planets (including the earth) formed by the accumulation of many bodies of smaller size. Surely among the multitudes of particles in the rings similar processes are being enacted on a small scale today. The ring systems thus offer a double challenge. One seeks to deduce the processes that made them and then one seeks to use this knowledge to gain insight into how the solid moons and planets formed.

THE FUTURE

During the 1970s, America's marvelous robot creations in space performed beyond all expectations, but Americans themselves became increasingly uncertain about their own global role and their long-range national objectives. Consequently, new investments were not made to maintain the rapid pace of planetary missions into the 1980s. The Soviet Union, however, continues to maintain a steady, long-term effort, exhibiting modest evolution in technologies, aimed especially at the planet Venus. As a consequence of these explorations, it can be expected that our understanding of Venus' surface, interior, and interaction between the surface and the atmosphere will increase. However, the Moon and Mars seem destined to continue to suffer from relative neglect following the Apollo and Viking missions of the 1960s and 1970s. Perhaps there will be one or two modest new automated ventures to either of them, although no country has announced any definite plans. Planet Mercury will remain entirely undisturbed by Earth's robots throughout this decade, at least.

Most significantly, deep space exploration will finally become genuinely international during the 1980s. European deep space efforts are planned, and Japan's first venture into the solar system is expected in the mid 1980s. In both cases, locally developed launch vehicles will be used for the first time to launch payloads beyond Earth orbit.

A most exciting event will be the space-borne explorations of Comet Halley during its 1986 reappearance in the inner solar system. This spectacular comet, with its well-documented 76-year period, will be investigated by two large Soviet spacecraft incorporating substantial French and Eastern European participation. These two VEGA (Venus–Halley) spacecraft will first be launched to Venus and then deflected on to Halley, where they will penetrate the bright fuzzy coma. In addition, the European Space Agency is developing a special spacecraft, designated *Giotto*. It will have a substantial dust shield to permit penetration of the dusty coma quite close to the nucleus itself in order to measure effectively the composition of materials ejected (as a consequence of the Sun's heating) from the rocky, icy mass that makes up the nucleus. Both the Soviet and European spacecraft will attempt to acquire images of the nucleus as well as the coma.

At the same time, a Japanese spacecraft will be passing the comet at greater distances than the other spacecraft. Its objective is to observe the ultraviolet radiation emitted and reflected by the coma, which should be of important supplementary scientific value. The United States will not be a participant in this first truly international exploration of an object in space.

The outer planets will continue to be explored by the Voyager 2 spacecraft as long as it remains functional. It is on a trajectory carrying it to an encounter with Uranus in 1986 and then on to the vicinity of Neptune in 1989. If successful, these additional encounters will provide substantially greater understanding of the outer solar system and will add significantly to the previous achievements of the Voyager program at Jupiter and Saturn. In addition, the United States plans to launch a sophisticated follow-up mission to Jupiter. The Galileo spacecraft will consist of a long-lived orbiter with instrumentation advanced beyond that carried on the Voyagers. Galileo will carry a sophisticated entry probe that will be dropped directly into the Jovian atmosphere to gather direct measurements of composition and other surface properties. Thus, knowledge of the outer planets will continue to increase as a result of American efforts lasting through the 1980s.

It becomes more difficult to forecast the character and magnitude of planetary exploration beyond the 1980s; the necessary preparatory efforts and critical national decisions still lie mostly in the future. However, it can be constructive to imagine some possible scenarios by which we may continue into the twenty-first century the brilliant beginning manifested in the Golden Decade of the 1970s.

First, it is always possible that the history of space exploration will repeat its previous pattern. There could be another nationalistic race similar to the one that spawned the Apollo program. However, when one considers the favorable circumstances leading to the Apollo race to the Moon, it seems unlikely that similar economic, political, and social conditions will recur to be the basis for a similar dramatic space race—this time with the goal being the first human visit to Mars. However, Mars plausibly could become the focus for a modest international program of automated spacecraft missions leading toward eventual human activity there well into the next century. In any case, foreseeable human endeavors at Mars will be of an adventuresome, rather than utilitarian, nature and thus can only occur when one or more nations is willing, for whatever reasons, to make substantial investments in a dramatic adventure in space.

Another possible scenario for planetary exploration is that this human endeavor will parallel the earlier exploration and eventual exploitation of the Antarctic continent. An early chauvinistic race to the South Pole (dominated more by personality than by national interest) occured at the beginning of this century. In the winter of 1911–1912, it was Scott against Amundsen, and Amundsen won. However, only limited exploration of the Antarctic continent occurred during the next four decades while Earth suffered two world wars. A by-product of those two wars was the development of comparatively inexpensive air and ocean transportation technology. Thus, after World War II, logistic support for Antarctic scientific exploration was affordable when considered within the total costs of maintaining navies and associated aircraft units. From such considerations, as well as from the entrepreneurial activities of scientific leaders in the United States and other countries, was born the International Geophysical Year (IGY) in 1957. IGY placed special emphasis on Antarctic studies, and even now many countries continue the modest programs of exploration and scientific measurements begun then.

However, the motivation for those activities has not been entirely altruistic. Almost all nations involved want to maintain a presence because of potential territorial claims. (The recent Falkland Islands conflict stands as a sharp reminder of such realities.) Resources in the Antarctic clearly will become valuable in the foreseeable future. The shrimplike krill, an important future source of food for a hungry Earth, are now being harvested in the Antarctic seas. There are also oil deposits, which will likely be exploited on a significant scale during coming decades.

So the Antarctic experience may provide a model for twenty-first century space exploration. The Apollo pattern may not be repeated. Rather, there may be an extended period of only modest deep space activities, such as seemingly will characterize the 1980s, during which space transportation technology will become less and less expensive, for reasons that have nothing to do with solar system exploration. Eventually it will become affordable to carry out new, ambitious (by current standards) exploration programs, probably designed with long-range utilitarian goals.

For example, visionaries for many years have regarded the Moon as a possible future locale for certain utilitarian activities. Certainly the far side of the Moon will be an important scientific region as it is forever free from the rapidly increasing radio frequency emanations from Earth and its orbiting satellites. The notion of exploitation of lunar materials for use there or in space, perhaps even in association with the capture of a small Earth-crossing asteroid, is science fiction now but may become reality a century from now.

Another utilitarian objective—perhaps even a twenty-first century global imperative—is detailed understanding of Earth's climatic changes. We are living on Earth during a brief episode of warm global climate in the midst of a long-term Ice Age that includes the entire history of human evolution and perhaps its future destiny as well. There is a tendency, perhaps because geologists refer to this period of Ice Ages as the *Pleistocene Period* and to the warm episode that still continues as the *Holocene*, to believe that the Ice Ages are over. However, there is no reason to suppose that is the case. Indeed, we not only face natural climatic changes in the future but also the global effects of large-scale combustion of fossil fuels and destruction of the world's forests. These human activities have added large amounts of carbon dioxide (CO_2) to our atmosphere. This growing CO_2 concentration in the atmosphere in turn is increasing the self-heating capacity of Earth, raising the prospect of seriously increased global temperature. Sustained higher temperatures would lead to partial melting of some ice caps and to the worldwide raising of sea levels. At the very least, serious local and regional climatic changes resulting from changed weather patterns are deemed likely by some specialists, even within the lifetimes of our children and grandchildren.

Thus, it may become most important to the citizens of Earth to obtain a detailed understanding of how the Sun's radiation interacts with Earth's climate so that scientists can better predict the nature of the global climatic change that Earth is likely to experience. A crucial part of this investigation must be carried out by specially constructed spacecraft located in various orbits about the Sun. In addition, Mars may contain clues to Earth's destiny. The polar areas of Mars probably contain very detailed sedimentary records of even minor and short-term variations in the Sun's luminosity. Unlike Earth, Mars has no oceans to stabilize its heat balance. Instead, it is characterized by annual global dust storms and polar deposition of water and CO_2 frost. Within a few decades, the full-scale exploration of Mars, especially its polar areas, could become an affordable utilitarian program of great significance.

If past patterns of exploration are maintained, the accomplishments of the Golden Decade and subsequent trends surely will lead to utilitarian activities in the next centuries that will prove most important to Earth's future inhabitants. We are just at the beginning of our attempt to explore and understand our solar system environment.

THE AUTHORS

CARL SAGAN is equally well known for his literary and for his scientific accomplishments. Winner of the Pulitzer Prize (for *The Dragons of Eden*), he has also received the NASA Medals for Exceptional Scientific Achievement and for Distinguished Public Service and the Joseph Priestley Award "for distinguished contributions to the welfare of mankind." He is David Duncan Professor of Astronomy and Space Sciences and Director of the Laboratory for Planetary Studies at Cornell University and has played a leading role in the Mariner, Viking, and Voyager expeditions to the planets. He has served as chairman of the Division for Planetary Sciences of the American Astronomical Society, president of the Planetology Section of the American Geophysical Union, and editor-in-chief of *Icarus*, the leading professional journal of solar system studies. A leader in establishing the high surface temperature of Venus and the existence of sand storms on Mars, Dr. Sagan was responsible for the Voyager interstellar record, a message about ourselves sent to other civilizations in space. His thirteen-part series COSMOS received the Peabody Award and became the most widely watched series in the history of American public television.

BRUCE MURRAY ("Mercury") is professor of planetary science at the California Institute of Technology and vice president of The Planetary Society. He was the Director of the Jet Propulsion Laboratory, which is operated by Cal Tech for the National Aeronautics and Space Administration, from 1976 to 1982. A geologist by training, he received his Ph.D.

from the Massachusetts Institute of Technology in 1955. Since going to Cal Tech in 1960 he has participated in five of the early Mariner missions, lastly as head of the group responsible for the Mariner 10 television experiment, which produced the first closeup pictures of Mercury and Venus.

GERALD SCHUBERT AND CURT COVEY ("The Atmosphere of Venus"): Schubert is currently a professor in the department of earth and space sciences and the department of geophysics and planetary physics at the University of California at Los Angeles. He was chairman of the Pioneer Venus work group on dynamics, wind, circulation, and turbulence from 1975 to 1981. He has been appointed editor of the *Journal of Geophysical Research* (Red) from October 1982 through 1986. He graduated from Cornell University in 1961 with degrees in engineering physics and aeronautical engineering. He then joined the Navy, serving for four years on the faculty of the Naval Nuclear Power School at Mare Island in California, while attending the University of California at Berkeley, where he obtained a Ph.D. in aeronautical sciences in 1964. After a postdoctoral year in the department of applied mathematics and theoretical physics at the University of Cambridge he returned to the U.S. to join the U.C.L.A. faculty in 1966.

Covey, a graduate student in the department of geophysics and planetary physics at U.C.L.A., majored in biology as an undergraduate at the Massachusetts Institute of Technology. He already has two advanced degrees, both from the

University of California at Santa Barbara: an M.A. in biology and an M.A. in physics.

GORDON H. PETTENGILL, DONALD B. CAMPBELL and HAROLD MASURSKY ("The Surface of Venus") have a common interest in the mapping of planetary surfaces. Pettengill is professor of planetary physics at the Massachusetts Institute of Technology. He did his undergraduate work at M.I.T. and his graduate work at the University of California at Berkeley, obtaining his Ph.D. from Berkeley in 1955. He then joined the staff of the Lincoln Laboratory at M.I.T., where he worked on the design of high-powered radars for the detection of intercontinental ballistic missiles. This work, he says, "led quite naturally into the field of radar astronomy." After serving for a time as director of the 300-meter radio telescope of the Arecibo Observatory in Puerto Rico, Pettengill returned to M.I.T. in 1970 as a member of the department of earth and planetary sciences.

Campbell is currently director of operations of the radar-astronomy group at the Arecibo Onbservatory. A native of Australia, he studied physics and radio astronomy at the University of Sydney before going to Arecibo in 1965 to work with Pettengill for a year on the newly completed radar system there. In 1966 Campbell went to Cornell University to complete work for his Ph.D., after which he returned to Arecibo.

Masursky is with the branch of astrogeological studies of the U.S. Geological Survey in Flagstaff, Arizona. He majored in geology as an undergraduate and graduate student at Yale University before joining the staff of the Geological Survey in 1951. Since then he has been involved in a variety of terrestrial and planetary research projects.

CONWAY B. LEOVY ("The Atmosphere of Mars") is professor of atmospheric sciences and geophysics and adjunct professor of astronomy at the University of Washington. Intrigued by the weather since boyhood, he first did his undergraduate work in physics at the University of Southern California and then joined the Air Force as an Air Weather Service forecaster and weather reconnaissance observer in the North Pacific and Central Pacific. After leaving the Air Force he went to the Massachusetts Institute of Technology, where he received his Ph.D. in meteorology in 1963. For the next five years he was a research meteorologist at the Rand Corporation, leaving in 1968 to go to the University of Washington. He has been interested in the atmospheres of the planets for many years; in 1968 he began working with Yale Mintz of the University of California at Los Angeles on models of the dynamics of the atmosphere of Mars. Since then he has been involved in the interpretation of the data on Martian atmospheric phenomena gathered by Mariner 6, Mariner 7, and Mariner 9 and in the meteorology experiments aboard the Viking landers.

RAYMOND E. ARVIDSON, ALAN B. BINDER and KENNETH L. JONES ("The Surface of Mars") are geologists with a special interest in the planets. Arvidson, who is associate professor of earth and planetary sciences at Washington University, was head of the Viking lander imaging team. He received his Ph.D. in geoscience from Brown University in 1974. His work on the surface of Mars, he writes, has "awakened an interest in the desert regions of the earth and how they recede and expand, particularly in regions where the spread of deserts is economically devastating." He is currently chairman of the data management computations committee for the Science Board of the National Academy of Space Sciences.

Binder, a research scientist at the Institute of Geophysics at the University of Kiel in West Germany, obtained his Ph.D. from the University of Arizona. He was involved in analyzing Viking and lunar sample data and "defending the theory of the formation of the moon by fission from the earth."

Jones, a research associate at Brown University, was a member of the Viking team. He obtained his Ph.D. from Brown University in 1974. He has recently founded a film production company specializing in scientific and technical documentaries.

ANDREW P. INGERSOLL ("Jupiter and Saturn") is professor of planetary science at the California Institute of Technology. He is a graduate of Amherst College (B.A., 1960) and Harvard University (M.A., 1961; Ph.D., 1965), and he has been at Cal Tech since 1966. A specialist in planetary atmospheres, he has taken part in the analysis of data from a number of U.S. space missions. He was awarded the National Aeronautics and Space Administration's Exceptional Scientific Achievement Medal for his work on the Voyager project. He writes: "I live with my wife, five children, several Cal Tech graduate students and a postdoc in a fine old Pasadena house that has sometimes been called 'the commune for hardworking scientists.' I enjoy challenges and am overcompetitive; I credit my family and friends for maintaining my civility."

LAURENCE A. SODERBLOM ("The Galilean Moons of Jupiter") is chief of the branch of astrogeologic studies of the U.S. Geological Survey. He has undergraduate degrees in both geology and physics from the New Mexico Institute of Mining and Technology and a doctorate in planetary science and geophysics from the California Institute of Technology. Since joining the Geological Survey in 1970 he has participated in several of the National Aeronautics and Space Administration's unmanned planetary-exploration missions. Soderblom is currently deputy team leader for the Voyager Imaging Science Experiment.

TOBIAS OWEN ("Titan") is professor of astronomy at the State University of New York at Stony Brook. His bachelor's and master's degrees, both in physics, are from the University of Chicago; he received his Ph.D. (in astronomy) from the University of Arizona in 1965. He has been a member of the imaging science team for Voyager 1 and Voyager 2. "My participation in the Voyager missions," he writes, "provided the totally unexpected opportunity to get close views and detailed information about the objects I had first known as points of light in the night skies of my childhood."

LAURENCE A SODERBLOM and TORRENCE V. JOHNSON ("The Moons of Saturn") are planetary scientists with a special interest in the satellites of the solar system. Soderblom is the author of "The Galilean Moons of Jupiter," which appeared in the January 1980 issue of SCIENTIFIC AMERICAN and is reprinted in this book.

Johnson teaches planetary science at Cal Tech and works as a senior research scientist at the Jet Propulsion Laboratory. A graduate of Washington University, he received his Ph.D. in planetary science from Cal Tech in 1970. He is currently a member of the Voyager imaging science team and a project scientist for the Galileo mission to Jupiter.

JAMES B. POLLACK and JEFFREY N. CUZZI ("Rings in the Solar System") work at the Ames Research Center of the National Aeronautics and Space Administration. Pollack is also chief scientist of the NASA Aerosol Climatic Effects Program. He earned his bachelor's and master's degrees in physics respectively at Princeton University in 1960 and the University of California at Berkeley in 1962; he received his Ph.D. in astronomy from Harvard University in 1965. Five years later he joined NASA, where he participated in the Pioneer Venus mission in 1978, the Mariner 9 mission in 1971, the Viking missions to Mars in 1976, and the Voyager mission to Saturn in 1980. Pollack's work has earned him two medals from NASA and one award from the American Institute of Aeronautics and Astronautics.

Cuzzi did his undergraduate work in engineering physics at Cornell University and obtained his Ph.D. in planetary science from the California Institute of Technology in 1972. Before going to Ames in 1974 he held research positions at the University of Massachusetts and at the Grand Ronde Center for Basic Studies. He writes: "I have studied the rings of Saturn and Jupiter for several years, both observationally and theoretically, most recently as a member of the Voyager imaging science team. I am particularly interested in what the dynamical behavior of rings may teach us about the formation of the planets themselves."

BIBLIOGRAPHIES

1. Mercury

SCIENCE, Vol. 195; July 12, 1974.

MAGNETIC FIELD OF MERCURY CONFIRMED. N. F. Ness, K. W. Behannon, R. P. Lepping and Y. C. Whang in *Nature*, Vol. 255, pages 204–205; May 15, 1975.

JOURNAL OF GEOPHYSICAL RESEARCH, Vol. 80, No. 17; June 10, 1975.

FLIGHT TO MERCURY. Bruce C. Murray and Eric Burgess. Columbia University Press, in press.

2. The Atmosphere of Venus

SCIENCE, Vol. 203, No. 4382; February 23, 1979. (Special issue devoted to Pioneer Venus.)

SCIENCE, Vol. 205, No. 4401; July 6, 1979. (Special issue devoted to Pioneer Venus.)

STRUCTURE AND CIRCULATION OF THE VENUS ATMOSPHERE. G. Schubert, C. Covey, A. D. Del Genio, L. S. Elson, G. Keating, A. Seiff, R. E. Young, J. Apt, C. C. Counselman III, A. J. Kliore, S. S. Limaye, H. E. Revercomb, L. A. Sromovsky, V. E. Suomi, F. Taylor, R. Woo and U. von Zahn in *Journal of Geophysical Research*, Vol. 85, pages 8007–8025; December, 1980.

THE THERMAL BALANCE OF VENUS IN LIGHT OF THE PIONEER VENUS MISSION. M. G. Tomasko, P. H. Smith, V. E. Suomi, L. A. Sromovsky, H. E. Revercomb, F. W. Taylor, D. J. Martonchik, A. Seiff, R. Boese, J. B. Pollack, A. P. Ingersoll, G. Schubert and C. C. Covey in *Journal of Geophysical Research*, Vol. 85, pages 8187–8199; December, 1980.

3. The Surface of Venus

MAPPING OF PLANETARY SURFACES BY RADAR. Tor Hagfors and Donald B. Campbell in *Proceedings of the IEEE*, Vol. 61, No. 9, pages 1219–1225; September, 1973.

PHYSICAL PROPERTIES OF THE PLANETS AND SATELLITES FROM RADAR OBSERVATIONS. G. H. Pettengill in *Annual Review of Astronomy and Astrophysics*, Vol. 16, pages 265–292; 1978.

4. The Atmosphere of Mars

MARINER 9 OBSERVATIONS OF THE MARS NORTH POLAR HOOD. Geoffrey A. Briggs and Conway B. Leovy in *Bulletin of the American Meteorological Society*, Vol. 55, No. 4, pages 278–296; April 1974.

COMPOSITION AND STRUCTURE OF THE MARTIAN ATMOSPHERE: PRELIMINARY RESULTS FROM VIKING 1. A. O. Nier, W. B. Hanson, A. Seiff, M. B. McElroy, N. W. Spencer, R. J. Duckett, T. C. D. Knight and W. S. Cook in *Science*, Vol. 193, No. 4255, pages 786–788; August 27, 1976.

ISOTOPIC COMPOSITION OF THE MARTIAN ATMOSPHERE. Alfred O. Nier, Michael B. McElroy and Yuk Ling Yung in *Science*, Vol. 194, No. 4060, pages 68–70; October 1, 1976.

STRUCTURE OF MARS' ATMOSPHERE UP TO 100 KILOMETERS FROM THE ENTRY MEASUREMENTS OF VIKING 2. Alvin Seiff and Donn B. Kirk in *Science*, Vol. 194, No. 4271, pages 1300–1302; December 17, 1976.

5. The Surface of Mars

MARS AS VIEWED BY MARINER 9: NASA SPECIAL PUBLICATION 329. National Aeronautics and Space Administration. Government Printing Office, 1974.

THE NEW MARS: THE DISCOVERIES OF MARINER 9. William K. Hartmann and Odell Raper. NASA Special Publication 337. Government Printing Office, 1974.

MARS. James B. Pollack in *Scientific American*, Vol. 233, No. 3, pages 107–117; September, 1975.

THE GEOLOGY OF MARS. Thomas A. Mutch, Raymond E. Arvidson, James W. Head, Kenneth L. Jones and R. Stephan Saunders. Princeton University Press, 1977.

SCIENTIFIC RESULTS OF THE VIKING PROJECT. *Journal of Geophysical Research*, Vol. 82, No. 28; September 30, 1977.

SURFACE OF MARS. Michael H. Carr. Yale University Press, 1981.

SCIENCE, Viking Issue, Vol. 194, December 17, 1976.

6. Jupiter and Saturn

NEGATIVE VISCOSITY. Victor P. Starr and Norman E. Gaut in *Scientific American*, Vol. 223, No. 1, pages 72–80; July, 1970.

THE METEOROLOGY OF JUPITER. Andrew P. Ingersoll in *Scientific American*, Vol. 234, No. 3, pages 46–56; March, 1976.

VOYAGER 1 ENCOUNTER WITH JUPITER. *Science*, Vol. 204, No. 4396; June 1, 1979.

VOYAGER 2 ENCOUNTER WITH JUPITER. *Science*, Vol. 206, No. 4421; November 23, 1979.

THE NEW SOLAR SYSTEM. Edited by J. Kelly Beatty, Brian O'Leary and Andrew Chaikin. Cambridge University Press and Sky Publishing Corporation, 1981.

VOYAGER 1 ENCOUNTER WITH SATURN. *Science*, Vol. 212, No. 4491; April 10, 1981.

VOYAGE TO JUPITER. David Morrison and Jane Samz. National Aeronautics and Space Administration, Publication SP0439.

SCIENCE, Voyager 1 Issue, Vol. 204, June 1, 1979.

SCIENCE, Voyager 2 Issue, Vol. 206, November 23, 1979.

NATURE, Voyager Issue, Vol. 280, August 30, 1979.

7. The Galilean Moons of Jupiter

MELTING OF IO BY TIDAL DISSIPATION. S. J. Peale, P. Cassen and R. T. Reynolds in *Science*, Vol. 203, No. 4383, pages 892–894; March 2, 1979.

INFRARED OBSERVATIONS OF THE JOVIAN SYSTEM FROM VOYAGER 1. R. Hanel, B. Conrath, M. Fraser, V. Kunde, P. Lowman, W. Maguire, J. Pearl, J. Pirraglia, R. Samuelson, D. Gautier, P. Gierasch, S. Kumar and C. Ponnamperuma in *Science*, Vol. 204, No. 4396, pages 972–976; June 1, 1979.

THE JUPITER SYSTEM THROUGH THE EYES OF VOYAGER 1. Bradford A. Smith, Laurence A. Soderblom, Torrence V. Johnson, Andrew P. Ingersoll, Stewart A. Collins, Eugene M. Shoemaker, G. E. Hunt, Harold Masursky, Michael H. Carr, Merton E. Davies, Allan F. Cook II, Joseph Boyce, G. Ed-

ward Danielson, Tobias Owen, Carl Sagan, Reta F. Beebe, Joseph Veverka, Robert G. Strom, John F. McCauley, David Morrison, Geoffrey A. Briggs and Verner E. Suomi in *Science*, Vol. 204, No. 4396, pages 951–972; June 1, 1979.

VOYAGER 1 ENCOUNTER WITH THE JOVIAN SYSTEM. E. C. Stone and A. L. Lane in *Science*, Vol. 204, No. 4396, pages 945–948; June 1, 1979.

VOYAGE TO JUPITER. David Morrison and Jane Samz. National Aeronautics and Space Administration, Publication SP0439.

SCIENCE, Voyager 1 Issue, Vol. 204, June 1, 1979.

SCIENCE, Voyager 2 Issue, Vol. 206, November 23, 1979.

NATURE, Voyager Issue, Vol. 280, August 30, 1979.

8. Titan

THE PLANET SATURN: A HISTORY OF OBSERVATION, THEORY AND DISCOVERY. A. F. O'D. Alexander. Faber and Faber, 1962.

TITAN. James B. Pollack in *The New Solar System*, edited by J. Kelly Beatty, Brian O'Leary and Andrew Chaikin. Cambridge University Press and Sky Publishing Corporation, 1981.

VOYAGER 1 ENCOUNTER WITH SATURN. *Science*, Vol. 212, No. 4491; April 10, 1981.

9. The Moons of Saturn

THE CHEMISTRY OF THE SOLAR SYSTEM. John S. Lewis in *Scientific American*, Vol. 230, No. 3, pages 50–65; March, 1974.

THE NEW SOLAR SYSTEM. Edited by J. Kelly Beatty, Brian O'Leary and Andrew Chaikin. Cambridge University Press and Sky Publishing Corporation, 1981.

VOYAGER 1 ENCOUNTER WITH SATURN. *Science*, Vol. 212, No. 4491; April 10, 1981.

A NEW LOOK AT THE SOLAR SYSTEM: THE VOYAGER 2 IMAGES. B. A. Smith, L. Soderblom, R. Batson, P. Bridges, J. Inge, M. Mazursky, E. Shoemaker, R. Beebe, J. Boyce, G. Briggs, A. Bunker, S. A. Collins, C. J. Kansen, T. Johnson, J. Mitchell, R. Terrile, A. F. Cook II, J. Cuzzi, J. Pollack, G. E. Danielson, A. Ingersoll, M. E. Davis, G. E. Hunt, D. Morrison, T. Owen, C. Sagan, J. Veverka, R. Strom and V. Suomi in *Science*, in press.

VOYAGES TO SATURN. David Morrison. National Aeronautics and Space Administration, Publication SP-451, 1982.

SCIENCE, Voyager 2 Issue, Vol. 215, January 29, 1982.

NATURE, Voyager 1 Issue, Vol. 292, August 20–26, 1981.

SCIENCE, Voyager 1 Issue, Vol. 212, April 10, 1981.

10. Rings in the Solar System

THE RINGS OF SATURN. James B. Pollack in *Space Sciences Review*, Vol. 18, No. 1, pages 3–93; October, 1975.

DISCOVERING THE RINGS OF URANUS. James L. Elliot, Edward Dunham and Robert L. Millis in *Sky and Telescope*, Vol. 53, No. 6, pages 412–416, 430; June, 1977.

JUPITER'S RINGS. Tobias Owen, G. Edward Danielson, Allan F. Cook, Candice Hansen, Virginia L. Hall and Thomas C. Duxbury in *Nature*, Vol. 281, No. 5731, pages 442–446; October 11, 1979.

ENCOUNTER WITH SATURN: VOYAGER 1 IMAGING SCIENCE RESULTS. Bradford A. Smith, Lawrence Soderblom, Reta Beebe, Joseph Boyce, Geoffrey Briggs, Anne Bunker, Stewart A. Collins, Candice J. Hansen, Torrence V. Johnson, Jim L. Mitchell, Richard J. Terrile, Michael Carr, Allen F. Cook II, Jeffrey Cuzzi, James B. Pollack, G. Edward Danielson, Andrew Ingersoll, Merton E. Davies, Garry E. Hunt, Arnold Masursky, Eugene Shoemaker, David Morrison, Tobias Owen, Carl Sagan, Joseph Veverka, Robert Strom and Verner E. Suomi in *Science*, Vol. 212, No. 4491, pages 163–191; April 10, 1981.

GENERAL BIBLIOGRAPHY

EARTHLIKE PLANETS. Bruce Murray, Michael C. Malin, Ronald Greeley. W. H. Freeman and Company, 1981.

COSMOS. Carl Sagan. Random House, 1980.

THE NEW SOLAR SYSTEM. Edited by J. Kelly Beatty, Brian O'Leary, and Andrew Chaikin. Sky Publishing Corporation and Cambridge University Press, 1981.

PLANETARY SCIENCE: A LUNAR PERSPECTIVE. Stuart Ross Taylor. Lunar and Planetary Institute, 1982.

INDEX